# Springer Theses

## Recognizing Outstanding Ph.D. Research

## Aims and Scope

The series "Springer Theses" brings together a selection of the very best Ph.D. theses from around the world and across the physical sciences. Nominated and endorsed by two recognized specialists, each published volume has been selected for its scientific excellence and the high impact of its contents for the pertinent field of research. For greater accessibility to non-specialists, the published versions include an extended introduction, as well as a foreword by the student's supervisor explaining the special relevance of the work for the field. As a whole, the series will provide a valuable resource both for newcomers to the research fields described, and for other scientists seeking detailed background information on special questions. Finally, it provides an accredited documentation of the valuable contributions made by today's younger generation of scientists.

## Theses are accepted into the series by invited nomination only and must fulfill all of the following criteria

- They must be written in good English.
- The topic should fall within the confines of Chemistry, Physics, Earth Sciences, Engineering and related interdisciplinary fields such as Materials, Nanoscience, Chemical Engineering, Complex Systems and Biophysics.
- The work reported in the thesis must represent a significant scientific advance.
- If the thesis includes previously published material, permission to reproduce this must be gained from the respective copyright holder.
- They must have been examined and passed during the 12 months prior to nomination.
- Each thesis should include a foreword by the supervisor outlining the significance of its content.
- The theses should have a clearly defined structure including an introduction accessible to scientists not expert in that particular field.

More information about this series at http://www.springer.com/series/8790

Boyang Shen

# Study of Second Generation High Temperature Superconductors: Electromagnetic Characteristics and AC Loss Analysis

Doctoral Thesis accepted by
the University of Cambridge, Cambridge, UK

 Springer

*Author*
Dr. Boyang Shen
Electrical Engineering Division
Department of Engineering
University of Cambridge
Cambridge, UK

*Supervisor*
Dr. Tim Coombs
Electrical Engineering Division
Department of Engineering
University of Cambridge
Cambridge, UK

ISSN 2190-5053        ISSN 2190-5061    (electronic)
Springer Theses
ISBN 978-3-030-58060-5        ISBN 978-3-030-58058-2    (eBook)
https://doi.org/10.1007/978-3-030-58058-2

This Springer imprint is published by the registered company Springer Nature Switzerland AG
The registered company address is: Gewerbestrasse 11, 6330 Cham, Switzerland

# Supervisor's Foreword

This thesis is a systematic study on the development of Second Generation (2G) High Temperature Superconductors (HTS). It includes a novel design of advanced medical imaging device using HTS, and an in-depth investigation on the losses of HTS.

The thesis demonstrates a considerable range of analytical and experimental research skills. It is built on the fundamental physics of superconductivity and considerably extends the body of knowledge on the behaviour of superconducting coils and tapes when subjected to AC fields and currents. This underpins a huge range of superconducting applications.

The thesis also describes the design and simulation of a superconducting Lorentz Force Electrical Impedance Tomography (LFEIT). This is a significant medical device that has the potential to be more efficient and compact than an MRI. It is designed to detect and enable the diagnosis of cancers at an early stage, as well as other pathologies such as stroke and internal haemorrhages, potentially saving hundreds and thousands of lives.

Overall, the research work carried out by author represents a significant contribution to the investigation of High Temperature Superconductors. This research is novel and central to the development of advanced superconducting applications in health care as well as more broadly in electrical and energy systems.

Cambridge, UK
May 2020

Dr. Tim Coombs

# Preface

Since the last century, superconducting applications have been making contributions to our everyday life, such as Magnetic Resonance Imaging (MRI) which uses strong magnetic fields and radio waves to produce detailed images of the body interior. In recent years, research on the electrical impedance of human tissues has become tremendously popular all over the world. The technologies to image the electrical impedance of biological tissues can make great contributions to the early diagnosis of cancer and other pathologies such as stroke, which could potentially save hundreds, even thousands, of lives.

Lorentz Force Electrical Impedance Tomography (LFEIT) is one of the most promising hybrid diagnostic scanners with burgeoning potential for biological imaging, particularly in the detection of cancer and internal haemorrhages. Superconducting magnet is one of the most important superconducting applications. Intense magnetic fields can be created when the coils are in the superconducting state, and they conduct a huge amount of electric current. The author tried a novel combination of superconducting magnets (based on the Second Generation (2G) High Temperature Superconductor (HTS)) together with the LFEIT system. Superconducting magnets can generate a magnetic field with high intensity and homogeneity, which could significantly enhance the electrical signal induced from a biological sample, thus improving the Signal-to-Noise Ratio (SNR), particularly for a full-body LFEIT.

Even if there are no actual alternating currents involved in the DC superconducting magnets for MRI and LFEIT, they have power dissipation during normal operation (e.g. magnet ramping) and under different alternating background fields. This problem generally goes under the category of "AC loss". Therefore, the AC loss characteristics of HTS tapes and coils are still fundamentally important for HTS magnet designs, even if they are normally operating in DC conditions. A comprehensive AC loss study on HTS tapes and HTS coils has been fulfilled, which is significantly crucial for the superconducting applications for transportation, health care, and electrical energy system.

The contents of this thesis have been highly recognised by the world top experts in superconductivity community. Based on the research achievements during my Ph.D. study, in 2018 I was awarded the "IEEE Council on Superconductivity Graduate Study Fellowship", which is the world highest prize for the graduates in superconductivity community (6 winners worldwide).

Cambridge, UK                                                                      Dr. Boyang Shen
May 2020

**Parts of this thesis have been published in the following peer-reviewed journal articles**

[1] **B. Shen**, C. Li, J. Geng, X. Zhang, J. Gawith, J. Ma, Y. Liu, F. Grilli and T. A. Coombs, "Power dissipation in HTS coated conductor coils under the simultaneous action of AC and DC currents and fields", *Superconductor Science and Technology*, vol. 31, no. 7, 2018.

[2] **B. Shen**, J. Li, J. Geng, L. Fu, X. Zhang, H. Zhang, C. Li, F. Grilli and T. A. Coombs, "Investigation of AC losses in horizontally parallel HTS tapes", *Superconductor Science and Technology*, vol. 30, no. 7, 2017.

[3] **B. Shen**, J. Geng, X. Zhang, L. Fu, C. Li, H. Zhang, Q. Dong, J. Ma, J. Gawith and T. A. Coombs, "AC losses in horizontally parallel HTS tapes for possible wireless power transfer applications", *Physica C: Superconductivity and Its Application*, vol. 543, pp. 35–40, 2017.

[4] **B. Shen**, J. Li, J. Geng, L. Fu, H. Zhang, X. Zhang, C. Li, Q. Dong, J. Ma and T. A. Coombs, "Investigation and Comparison of AC Losses on Stabilizer-free and Copper Stabilizer HTS Tapes", *Physica C: Superconductivity and Its Application*, vol. 541, pp. 40–44, 2017.

[5] **B. Shen**, J. Geng, C. Li, X. Zhang, L. Fu, H. Zhang, J. Ma and T. A. Coombs, "Optimization Study on the Magnetic Field of Superconducting Halbach Array Magnet", *Physica C: Superconductivity and Its Application*, vol. 538, pp. 46–51, 2017.

[6] **B. Shen**, L. Fu, J. Geng, H. Zhang, X. Zhang, Z. Zhong, Z. Huang and T. A. Coombs, "Design of a Superconducting Magnet for Lorentz Force Electrical Impedance Tomography", *IEEE Transactions on Applied Superconductivity*, vol. 26, no. 3, 2016.

[7] **B. Shen**, L. Fu, J. Geng, X. Zhang, H. Zhang, Q. Dong, C. Li, J. Li and T. A. Coombs, "Design and Simulation of Superconducting Lorentz Force Electrical Impedance Tomography (LFEIT)", *Physica C: Superconductivity and Its Application*, vol. 524, pp. 5–12, 2016.

# Acknowledgements

First of all, I am particularly grateful to my supervisor Dr. Tim Coombs. His valuable comments and guidance have provided me with significant help throughout my Ph.D. study at the University of Cambridge. Without his great support, my Ph.D. work would not have proceeded. I would like to express my sincere gratitude to Dr. Coombs's wife, Ching-fen Coombs, for taking care of me during my Ph.D. study at Cambridge.

I would like to express my tremendous gratitude to Dr. Francesco Grilli for his precious advice and crucial support in the modelling of superconductivity since the first time we met at the HTS modelling school at École Polytechnique Fédérale de Lausanne (EPFL), Switzerland, in June 2016. I would like to thank Prof. Mathias Noe and Dr. Francesco Grilli for their invitation to Karlsruhe Institute of Technology (KIT), Germany, in September 2017. As a "Guest Scientist" at KIT, I did interesting research, learned useful knowledge, and met brilliant people.

I would like to say thanks a lot to my advisor Prof. Tim Wilkinson. His important comments and instructions have provided me with great help for my study.

I also would like to say a big thank you to my colleagues Dr. Harold Ruiz, Dr. Koichi Matsuda, Dr. Wei Wang, Dr. Mehdi Baghdadi, Dr. Zhaoyang Zhong, Dr. Zhen Huang, Dr. Yujia Zhai, Dr. Jing Li, Dr. Lin Fu, Jianzhao Geng, Heng Zhang, Xiuchang Zhang, Qihuan Dong, Chao Li, Jun Ma, James Gawith, Kaihe Zhang, and Jiabin Yang. Their generous help ensured the success of this work.

The experimental work was carried out with the assistance from the Electrical Engineering Division, Department of Engineering, University of Cambridge. I would like to express huge gratitude to Mr. John Grundy and other members of staff for their important help.

I am grateful to Wolfson College and the Department of Engineering, University of Cambridge, for their academic travel/conference grants, which helped me to travel worldwide in the pursuit of more and more knowledge.

Finally, I would like to express great appreciation to my parents, my wife, and family members for their understanding, continuous support, and endless love.

# Contents

# Abbreviations

| | |
|---|---|
| 1D | One-dimension |
| 2D | Two-dimension |
| 3D | Three-dimension |
| 1G | First Generation |
| 2G | Second Generation |
| $\hbar$ | Planck Constant |
| $\lambda$ | Penetration Depth |
| $\xi$ | Coherence Length |
| $\phi$ | Magnetic Flux |
| $\mu$ | Permeability |
| $\rho$ *(mass)* | Mass Density |
| $\rho$ *(resistivity)* | Electrical Resistivity |
| AC | Alternating Current |
| $B$ | Magnetic Flux Density |
| BSCCO | $Bi_2Sr_2CaCu_2O_8$ |
| $C$ | Capacitance |
| CMOS | Complementary Metal Oxide Semiconductor |
| dB | Decibels |
| DC | Direct Current |
| DSP | Digital Signal Processor |
| DSV | Diameter of Spherical Volume |
| $E$ | Electric Field |
| $f$ | Frequency |
| FEM | Finite Element Method |
| G | Giga |
| GL | Ginzburg-Landau |
| $H$ | Magnetic Field Intensity |
| HEI | Hall Effect Imaging |
| HTS | High Temperature Superconductor |
| Hz | Hertz |

| $I$ | Current |
|---|---|
| $J$ | Current Density |
| k | Kilo |
| K | Kelvin |
| $L$ | Inductance |
| LFEIT | Lorentz Force Electrical Impedance Tomography |
| LTS | Low Temperature Superconductor |
| $m$ | mass |
| M | Mega |
| MAET | Magneto-Acousto-Electric Tomography |
| MAT-MI | Magneto-acoustic Tomography with Magnetic Induction |
| MRI | Magnetic Resonance imaging |
| NI | National Instruments |
| ppk | Parts per Thousand |
| ppm | Parts per Million |
| PDE | Partial Differential Equation |
| $Q$ | Energy (Joule) |
| $R$ | Resistance |
| RMS | Root Mean Square |
| SCS | Surround Copper Stabilizer |
| SF | Stabilizer-free |
| SFCL | Superconducting Fault Current Limiter |
| SNR | Signal to Noise Ratio |
| $t$ | Time |
| $Tc$ | Critical Temperature |
| $T$ | Tesla |
| $V$ | Voltage |
| $w$ | Width |
| $W$ | Power (Watt) |
| WPT | Wireless Power Transfer |
| YBCO | $YBa_2Cu_3O_7$ |
| $Z$ | Impedance |

# List of Figures

# List of Tables

# Chapter 1
# Introduction

## 1.1 Motivation

Since the last century, superconducting applications have been making contributions to our everyday life, such as Magnetic Resonance Imaging (MRI) which uses strong magnetic fields and radio waves to produce detailed images of the body interior [1]. In recent years, research on the electrical impedance of human tissues has become tremendously popular all over the world [2]. The technologies to image the electrical impedance of biological tissues can make great contributions to the early diagnosis of cancer and other pathologies such as stroke, which could potentially save hundreds, even thousands, of lives [3]. Lorentz Force Electrical Impedance Tomography (LFEIT) is a novel and promising configuration for diagnostic scanners. The working principle of LFEIT is to image the electrical signal generated by the magneto-acoustic effect from a biological sample [4]. LFEIT can achieve 3D high resolution imaging of tissue impedance based on an ultrasonically induced Lorentz force [3, 5–7].

One of the most important superconducting applications is the superconducting magnet [8]. A superconducting magnet is an electromagnet made from superconducting coils. Intense magnetic fields can be created when the coils are in the superconducting state, and they conduct a huge amount of electric current [8, 9]. The first successful superconducting magnet was built by George Yntema in 1954 using niobium wire, and achieved a field of 0.71 T at 4.2 K [10]. In 2017 a superconducting magnet achieved the new world record of a 32 T magnetic field at the National High Magnetic Field Laboratory, USA [11].

For both MRI and LFEIT, magnets are the key components. The intensity and the uniformity of magnetic field directly affect the imaging quality of biological tissues [3, 5]. Therefore, the combination of superconducting magnet with LFEIT system

B. Shen, *Study of Second Generation High Temperature Superconductors:
Electromagnetic Characteristics and AC Loss Analysis*, Springer Theses,
https://doi.org/10.1007/978-3-030-58058-2_1

could be a sensible strategy. A superconducting magnet is able to produce a magnetic field with high intensity and uniformity, which could be significantly beneficial for the enhancement of electrical signal output and the robustness to noise, especially for large scale LFEIT systems.

For Direct Current (DC) systems, theoretically, superconductors present the electrically lossless attribute in most conditions [12]. The superconducting magnets are generally operating in the DC condition for MRI and LFEIT. Nevertheless, when high current superconducting coils and cables are used in magnet applications, they dissipate power because they are exposed to the varying magnetic field generated by the magnet itself, as well as to the ambient factors such as the external AC signal disturbances. These kinds of power dissipation are mainly due to the magnetisation loss and dynamic loss from superconducting materials. Although no actual alternating currents are involved, these sorts of problems are generally classified under the category of "AC loss", because the dynamics of power dissipation during a field ramp are substantially the same as the problems encountered in AC conditions. Field ramping and other transient changes of the transport current generate heat and challenge the cryogenic systems. A small AC magnetic field disturbance of 10 mT at kilo hertz level can affect the stability of a superconducting magnet [13]. Therefore, it is crucial to investigate the AC loss characteristics of HTS tapes and coils for the design of superconducting magnets, even if these magnets are normally operating in DC conditions.

## 1.2 Thesis Novelties

There are 7 main novelties which are located in 6 main chapters of this thesis. These novelties have been published in 7 peer-reviewed journal articles listed below:

(1) Design of a Superconducting Magnet for Lorentz Force Electrical Impedance Tomography [14].
(2) Optimization Study on the Magnetic Field of Superconducting Halbach Array Magnet [15].
(3) Design and Simulation of Superconducting Lorentz Force Electrical Impedance Tomography (LFEIT) [16].
(4) Investigation and comparison of AC losses on Stabilizer-free and Copper Stabilizer HTS tapes [17].
(5) Investigation of AC losses in horizontally parallel HTS tapes [18].
(6) AC losses in horizontally parallel HTS tapes for possible wireless power transfer applications [19].
(7) Power dissipation in HTS coated conductor coils under the simultaneous action of AC and DC currents and fields [20].

## 1.3  Thesis Outline

This chapter introduces the motivation, novelties and outline of this thesis.

Chapter 2 presents a literature review of superconductivity, the basics of AC loss and the historical development of LFEIT. The review of superconductivity covers the Meissner Effect, the London Theory, the Ginzburg-Landau Theory, Low $T_c$ and High $T_c$ Superconductors, Type I and Type II Superconductors, BCS Theory, and Flux Pinning. The basics of AC loss consist of the AC loss classifications as well as their origins, and AC loss measurement methods.

Chapter 3 describes the numerical solutions and theoretical analysis of this design, and is in two parts. The first part covers the modelling and simulation of High $T_c$ Superconductors, including the Critical State Model, the Bean Model, the Kim model, the $E$-$J$ Constitutive Power Law, the General H-formulation, Two-dimension (2D) H-formulation Models, and Three-dimension (3D) H-formulation Models. The second part is a theoretical analysis of LFEIT which includes the acoustic field study and the magneto-acousto-electric technique.

Chapter 4 demonstrates four magnet designs for the LEFIT system, which are: (1) Halbach Array magnet design (perfect round shaped permanent magnets), (2) Halbach Array magnet design (square shaped permanent magnets), (3) Superconducting Helmholtz Pair magnet design, and (4) Superconducting Halbach Array magnet design. The working principles and design specifications of these four magnets are described. The modelling and simulation of these magnets were based on COMSOL Multiphysics. The permanent magnets design used the "Magnetic field (mf)" AC/DC module while the superconducting magnets design used the Partial Differential Equation (PED) model. More detailed analysis on the magnitude and uniformity of the magnetic field for the Superconducting Halbach Array magnet design is also presented. Finally, the advantages and shortcomings of the four magnet designs are discussed.

Chapter 5 illustrates the optimization study of the superconducting Halbach Array magnet carried out using FEM methods based on COMSOL Multiphysics, with 2D models using $H$-formulation based on the $B$-dependent critical current density and bulk approximation. The optimization dedicates the location and geometry of HTS coils, and numbers of coils without changing the total amount of superconducting material used. Mathematical relations were developed for these optimization parameters involving the intensity and homogeneity of the magnetic field, for the purpose of predicting optimization performance and efficiency.

Chapter 6 presents the design of the LFEIT system, which includes the design of the ultrasound module and magnetic module for LFITE system, and the electrical signal simulation based on the mathematical model of LFEIT. The mathematical model simulated two samples tested by this LFEIT system, where they were located in three magnetic fields with different magnetic strengths and uniformity. Then the comparison and discussion of their electrical signal outputs and basic signal imaging are presented.

   Chapter 7 clears the necessity of AC loss study for superconducting magnet designs, as well as their connections. It starts with the investigation of AC losses in Surround Copper Stabilizer (SCS) Tape and Stabilizer-free (SF) Tape, which includes AC loss measurement using the electrical method, and the real geometry and multi-layer HTS tape modelling using the $H$–formulation by COMSOL. Hysteresis AC losses in the superconducting layer, and eddy–current AC losses in the copper stabilizer, silver overlayer and substrate are the focus on this study. The measured AC losses were compared to the AC losses from the simulation, for three cases of different AC frequency: 10 Hz, 100 Hz, and 1000 Hz. The eddy-current AC losses of copper stabilizer at a frequency of 1000 Hz were determined from both experiment and simulation. The frequency dependence of AC losses from Stabilizer–free Tape and Copper Stabilizer Tape are compared and analysed.

   Chapter 8 demonstrates a comprehensive AC loss study of a circular HTS coil. The AC losses from a circular double pancake coil were measured using the electrical method. A 2D axisymmetric $H$-formulation model using the FEM package COMSOL Multiphysics has been established, which was able to make consistency with the real circular coil used in the experiment. Three scenarios are analysed: (1) AC transport current and DC magnetic field, (2) DC transport current and AC magnetic field, and (3) AC transport current and AC magnetic field. The angular dependence analysis of the coil under the magnetic field with the different orientation angle $\theta$ has been performed for all three scenarios. In scenario (3), the effect of relative phase difference $\Delta\varphi$ between the AC current and the AC field on the total AC loss of the coil is studied. In short, a current/field/angle/phase dependent AC loss $(I, B, \theta, \Delta\varphi)$ study of circular HTS coil has been fulfilled, which could be helpful for future design and research of HTS AC systems.

   Chapter 9 presents the AC losses of horizontally parallel HTS tapes. A Three-tape configuration has been set up as an example. By using the electrical method, the AC losses of the middle and end tapes of three parallel tapes have been measured and compared to those of an individual isolated tape. The effect of the interaction between tapes on AC losses is analysed, and compared with simulations using the 2D $H$–formulation by COMSOL. The electromagnetic induction and AC losses generated by a conventional 3-turn coil was calculated, and then compared to the case of three parallel tapes with the same condition. The analysis reveals that HTS parallel tapes could be potentially applied into wireless power transfer (WPT) systems, which could suffer lower total AC losses than general HTS coils. The cases of increasing numbers of parallel tapes have been proposed, and the normalised ratio between the total average AC losses per tape and the AC losses of an individual single tape has been computed for different gap distances, by using the FEM calculation. A new parameter has been proposed, $N_s$, a transition point for the number of tapes, to divide Stage 1 and Stage 2 for the AC loss study of horizontally parallel tapes. For Stage 1, $N < N_s$, the total average losses per tape increased with the increasing number of tapes. For Stage 2, $N > N_s$, the total average losses per tape started to decrease with the increasing number of tapes. The results proved that horizontally parallel HTS tapes could be potentially utilised for superconducting devices such as HTS transformers, which could maintain or even reduce the total average AC losses per tape using huge numbers of parallel tapes.

Chapter 10 summarises the conclusions of the work completed and makes suggestions for future work.

# References

1. R.R. Edelman, S. Warach, Magnetic resonance imaging. N. Engl. J. Med. **328**(11), 785–791 (1993)
2. Y. Zou, Z. Guo, A review of electrical impedance techniques for breast cancer detection. Med. Eng. & Phys. **25**(2), 79–90 (2003)
3. N. Polydorides, *In-Vivo Imaging With Lorentz Force Electrical Impedance Tomography* (University of Edinburgh, UK, 2014)
4. H. Wen, J. Shah, R.S. Balaban, Hall effect imaging. IEEE Trans. Biomed. Eng. **45**(1), 119–124 (1998)
5. P. Grasland-Mongrain, J. Mari, J. Chapelon, C. Lafon, Lorentz force electrical impedance tomography. IRBM **34**(4), 357–360 (2013)
6. S. Haider, A. Hrbek, Y. Xu, Magneto-acousto-electrical tomography: a potential method for imaging current density and electrical impedance. Physiol. Meas. **29**(6), 41–50 (2008)
7. A. Montalibet, J. Jossinet, A. Matias, D. Cathignol, Electric current generated by ultrasonically induced Lorentz force in biological media. Med. Biol. Eng. Compu. **39**(1), 15–20 (2001)
8. H. Brechna, *Superconducting Magnet Systems* (1973)
9. A.C. Rose-Innes, *Introduction to superconductivity*: Elsevier (2012)
10. M.N. Wilson, 100 years of superconductivity and 50 years of superconducting magnets, Appl. Supercond. IEEE Trans. **22**(3) (2012), Art. no. 3800212
11. *The World's Strongest Superconducting Magnet Breaks a New Record*, National High Magnetic Field Laboratory (2017)
12. F. Grilli, E. Pardo, A. Stenvall, D.N. Nguyen, W. Yuan, F. Gömöry, Computation of losses in HTS under the action of varying magnetic fields and currents. IEEE Trans. Appl. Supercond. **24**(1), 78–110 (2014)
13. Y. Iwasa, *Case studies in superconducting magnets: design and operational issues*: Springer Science & Business Media (2009)
14. B. Shen, L. Fu, J. Geng, H. Zhang, X. Zhang, Z. Zhong, Z. Huang, T. Coombs, Design of a superconducting magnet for lorentz force electrical impedance tomography, IEEE Trans. Appl. Supercond. **26**(3) (2016), Art. no. 4400205
15. B. Shen, J. Geng, C. Li, X. Zhang, L. Fu, H. Zhang, J. Ma, T. Coombs, Optimization study on the magnetic field of superconducting Halbach Array magnet. Phys. C, Supercond. **538**, 46–51 (2017)
16. B. Shen, L. Fu, J. Geng, X. Zhang, H. Zhang, Q. Dong, C. Li, J. Li, T. Coombs, Design and simulation of superconducting Lorentz Force Electrical Impedance Tomography (LFEIT). Phys. C, Supercond. **524**, 5–12 (2016)
17. B. Shen, J. Li, J. Geng, L. Fu, X. Zhang, C. Li, H. Zhang, Q. Dong, J. Ma, T.A. Coombs, Investigation and comparison of AC losses on stabilizer-free and copper stabilizer HTS tapes. Phys. C, Supercond. **541**, 40–44 (2017)
18. B. Shen, J. Li, J. Geng, L. Fu, X. Zhang, H. Zhang, C. Li, F. Grilli, T.A. Coombs, Investigation of AC losses in horizontally parallel HTS tapes, Supercond. Sci. Technol. **30**(7) (2017), Art. no. 075006

19. B. Shen, J. Geng, X. Zhang, L. Fu, C. Li, H. Zhang, Q. Dong, J. Ma, J. Gawith, T.A. Coombs, AC losses in horizontally parallel HTS tapes for possible wireless power transfer applications. Phys. C, Supercond. **543**, 35–40 (2017)
20. B. Shen, C. Li, J. Geng, X. Zhang, J. Gawith, J. Ma, Y. Liu, F. Grilli, T A. Coombs, Power dissipation in HTS coated conductor coils under the simultaneous action of AC and DC currents and fields, Supercond. Sci. Technol. **31**(7) (2018)

# Chapter 2
# Literature Review

## 2.1 Review of Superconductivity

Superconductivity was first discovered by Heike Kamerlingh-Onnes on the 8th of April 1911 [1]. At that time, he was studying the resistance of solid mercury at cryogenic temperatures using liquid helium as a refrigerant, and at 4.2 K he observed that its resistance abruptly disappeared [1, 2]. From then on, superconductivity has aroused great interest, providing solutions to many challenges. Huge progress has been made in current engineering applications such as superconducting motors, superconducting transmission lines, superconducting fault current limiters [1, 3].

### 2.1.1 Meissner Effect

In 1933, 22 years after Kamerlingh-Onnes's discovery of superconductivity, the Meissner effect phenomenon was discovered by Walther Meissner and Robert Ochsenfeld [4]. The Meissner effect describes the phenomenon: Metal in the superconducting state never allows a magnetic flux density to exist in its interior [4]. The Meissner effect differs from the behaviour of a theoretical perfect conductor ($\sigma = \infty$).

Detailed comparisons between a perfect conductor and a superconductor are shown in Fig. 2.1. Figure 2.1 (Red box) presents the magnetic behaviour of a perfect conductor, while Fig. 2.1 (Green box) demonstrates the magnetic behaviour of a superconductor: From (a)–(b) Specimen is resistanceless in the absence of a field; (c) Magnetic field is applied to the specimen; (d) Magnetic field is removed; (e)–(f) Specimen is resistanceless in an applied magnetic field; (g) Applied magnetic field is removed. It can be seen that when the specimen is cooled before applying the magnetic field (zero field cooling), both a perfect conductor and a superconductor

© The Editor(s) (if applicable) and The Author(s), under exclusive license
to Springer Nature Switzerland AG 2020
B. Shen, *Study of Second Generation High Temperature Superconductors: Electromagnetic Characteristics and AC Loss Analysis*, Springer Theses,
https://doi.org/10.1007/978-3-030-58058-2_2

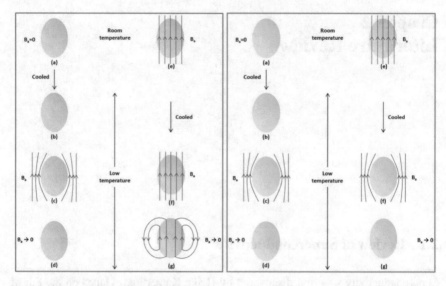

**Fig. 2.1** (Red box) Magnetic behaviour—a perfect conductor. (Green box) Magnetic behaviour—a superconductor

are able to expel the magnetic field from the inner geometry [5]. On the contrary, if the magnetic field is applied before cooling the specimen, for an ideal conductor the magnetic field remains within the geometry after it has been cooled down, whereas for a superconductor the magnetic field is still expelled from the inner geometry after the sample is cooled below its $T_c$. The expulsion of a magnetic field in a super-conductor is due to the Meissner effect, while the magnetic expulsion in a perfect conductor is caused by Lenz's law [5].

According to the phenomenon of the Meissner Effect that a magnetic field is expelled from the body of a superconductor, it can be deduced:

$$B = \mu_0(H + M) = 0 \tag{2.1}$$

where $B$ is the magnetic flux density, $\mu_0$ is the permeability of free space, $H$ is the magnetic field intensity, and $M$ is the magnetisation. Equation (2.1) gives:

$$H = -M \tag{2.2}$$

Therefore, the susceptibility:

$$\chi = \frac{dM}{dH} = -1 \tag{2.3}$$

Assuming the ideal resistivity for a superconductor:

$$\rho = 0 \tag{2.4}$$

The diamagnetism of a superconductor arises from the shielding current at its surface. This shielding current induces a magnetic field that has the same magnitude yet the opposite direction as the external field [5]. As a result, the sum of these two magnetic fields can be exactly cancelled. Therefore, expulsion of the total magnetic field can be seen and there is a zero magnetic field inside the superconductor [5].

## 2.1.2 The London Theory

Since the surface current density induced by super electrons is not infinitely large, the magnetic field that leads to ideal diamagnetism of a superconductor must penetrate the superconductor into a finite depth [5, 6]. The penetration depth $\lambda$ can be derived from basic electrodynamics characteristics, and in the following paragraphs the phenomenological London Theory is explained according to the penetration depth $\lambda$ varying with the super current density $n_s$ [5, 6].

Superelectrons encounter zero resistance to their motion in a superconductor. This assumes that near the surface of the superconductor, a constant electric field $E$ exists in the material, and the electrons accelerate steadily under the action of this electric field.

$$m_s \dot{v} = Eq \tag{2.5}$$

where $m_s$ is the mass of the free carrier, $v$ is the carrier velocity and $q$ is the carrier charge (superelectron or copper-pairs), and the time derivative is indicated by using the dot on the top (e.g. the accelerator $\dot{v}$) [6]. The super current density can then be expressed as:

$$J = n_s v q \tag{2.6}$$

where $n_s$ is the superelectrons per unit volume. In 1925, Haas Lorentz predicted the electrodynamic screening length $\lambda$ to be [7]:

$$\lambda^2 = \frac{m_s}{\mu_0 n_s q^2} \tag{2.7}$$

Rearrange the (2.5), (2.6) and (2.7):

$$E = \frac{m_s \dot{v}}{q} = \frac{m_s \dot{J}}{n_s q^2} = \mu_0 \lambda^2 \dot{J} \tag{2.8}$$

where $\overset{\bullet}{J}$ is the time derivative of the current density. For both sides of Eq. (2.8), taking the curl:

$$\nabla \times E = \mu_0 \lambda^2 \nabla \times \overset{\bullet}{J} \tag{2.9}$$

Using Maxwell Equations:

$$\nabla \times E = -\overset{\bullet}{B} \tag{2.10}$$

$$\nabla \times B = \mu_0 J \tag{2.11}$$

where $\overset{\bullet}{B}$ is the time derivative of the magnetic flux density, Eq. (2.9) can be rearranged:

$$\overset{\bullet}{B} + \lambda^2 \nabla \times \nabla \times \overset{\bullet}{B} = 0 \tag{2.12}$$

As the second term:

$$\nabla \times \nabla \times \overset{\bullet}{B} = -\nabla^2 \overset{\bullet}{B} \tag{2.13}$$

Equation (2.12) can be converted:

$$\overset{\bullet}{B} = \lambda^2 \nabla^2 \overset{\bullet}{B} \tag{2.14}$$

The solution of Eq. (2.14) must satisfy the condition of diamagnetism:

$$\overset{\bullet}{B}(z) = \overset{\bullet}{B}_a e^{\left(-\frac{z}{\lambda}\right)} \tag{2.15}$$

where $B_a$ is the external applied magnetic flux, $B_a = \mu_0 H_a$.

From Eq. (2.15), the condition for a perfect conductor is $B(z) = 0$ when $z >> \lambda$. However, a superconductor is more than a perfect conductor [8]. H. London and F. London performed the time integration of Eq. (2.15) then converted this equation with a static magnetic flux $B$, and set the additional unknown integration constant to be equal to zero. They discovered that [9, 10]:

$$B(z) = B_a e^{\left(-\frac{z}{\lambda}\right)} \tag{2.16}$$

Figure 2.2 shows how magnetic field penetrates into a superconductor according to the London Theory. Within a distance $\lambda$ near the surface, the magnetic field decays exponentially, but the deep interior of the superconductor is free from any magnetic field [8, 9].

**Fig. 2.2** London penetration
depth: the penetration of a
magnetic field into a
superconductor [9]

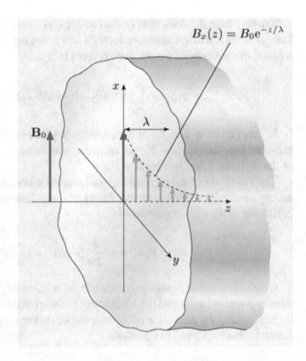

The London Theory uses a mathematical way to explain the Meissner Effect [5, 9]. The solution to Eq. (2.16) precisely describes the phenomenon of the Meissner Effect: when a superconductor is surrounded by an external static magnetic field, supercurrents arise spontaneously in the surface layer with a penetration depth of $\lambda$, creating an opposing field in the sample, which can exactly cancel the applied field inside [5, 9].

### 2.1.3  The Ginzburg-Landau (GL) Theory

In 1950, Ginzburg and Landau proposed the Ginzburg-Landau Theory, which is of paramount importance in the history of superconductivity. The Ginzburg-Landau Theory is a phenomenological explanation of superconductivity in terms of thermodynamics [10, 11]. The GL Theory describes the long coherence length for some superconductors, e.g. low temperature superconductors (LTS). It also explains the short coherence length for high temperature superconductors (HTS) [11–13].

The Ginzburg-Landau Theory starts from some simple assumptions, which explain superconductivity based on a complex order parameter $\phi$, and it can be expressed in the form of the products which include a modulus $|\phi|$ and a phase factor $\theta$:

$$\phi = \sqrt{n_s}\exp(i\theta) \tag{2.17}$$

Starting from the deduction of the Landau free energy density, both Ginzburg and Landau assumed that a superconductor's free energy density $F_s$ can be expanded as a function with complex order parameters [11]:

$$F_s = F_n + \alpha |\phi|^2 + \frac{\beta}{2} |\phi|^4 + \frac{1}{2m} |(-i\hbar \nabla - 2eA)\phi|^2 + \frac{B^2}{2\mu_0} \tag{2.18}$$

In Eq. (2.18), $\alpha$ and $\beta$ are temperature dependent functions, $A$ is the magnetic potential (vector), $\hbar$ is the Planck constant, and $e$ and $m$ are the charge and the mass of the particles described by $\phi$.

The variations of the parameter $\phi$ and magnetic potential $A$ which lead to the free energy must be at a minimum. The First Ginzburg-Landau equation uses the varying derivative of $F_s$ with respect to $\phi$:

$$\alpha\phi + \beta |\phi|^2 \phi + \frac{1}{2m} (-i\hbar \nabla - 2eA)^2 \phi = 0 \tag{2.19}$$

Ginzburg-Landau Equation I describes the relationship between the superelectron density $\phi$ and the magnetic field. Similarly, to determine the expression for the current density, taking the derivative of $F_s$ with respect to the magnetic potential $A$ gives the Second Ginzburg-Landau Equation:

$$J = \frac{e}{m} [\phi^*(-i\hbar \nabla - 2eA)\phi] + c.c. \tag{2.20}$$

The Ginzburg-Landau Equations are able to produce many useful results [13]. The most impactful contribution is their prediction of the existence of two characteristic lengths for a superconductor [8, 11]. The first one is the coherence length $\xi$, which explains the size of thermodynamic fluctuations during the superconducting $(T < T_c)$ phase:

$$\xi = \sqrt{\frac{\hbar^2}{4m\alpha(T)}} \quad (T < T_c) \tag{2.21}$$

When $T > T_c$ (normal phase), it is given by:

$$\xi = \sqrt{\frac{\hbar^2}{2m\alpha(T)}} \quad (T > T_c) \tag{2.22}$$

The second prediction is the penetration depth $\lambda$, given by:

$$\lambda = \sqrt{\frac{m}{4\mu_0 e^2 \phi^2}} \tag{2.23}$$

As mentioned before, $\phi$ is the equilibrium quantity to the order parameter, which is in the absence of electromagnetic field. The penetration depth $\lambda$ describes the depth to which an external magnetic field can penetrate into a superconductor [11, 12].

### 2.1.4 Low $T_c$ and High $T_c$ Superconductors

Since Kamerlingh-Onnes's first discovery of superconductor, a great amount of research work has been carried out to find more and more superconductors, and scientists are always searching for higher critical temperature ($T_c$) superconductors [14]. Thereafter, low $T_c$ and high $T_c$ superconductors were distinguished according to their critical temperatures [1, 14]. The earliest discovered superconducting materials were mostly metals and simple alloys. For pure metal, Niobium is the element has the highest critical temperature (9.3 K), which generally requires liquid helium (He) as the cryogen [1, 15]. However, some metallic compounds and alloys remain superconducting state up to much higher temperatures. The $T_c$ of Lanthanum-barium-copper oxide ceramic ($LaBaCuO_4$) was measured at 30 K, which was used as the criterion to define high temperature superconductors (HTS). Since then, superconductors have been classified into 2 categories [2, 16] (Table 2.1).

In recent years, a great number of high temperature superconductors (HTS) based on copper oxide have been discovered, and are used for research and engineering applications, e.g. $YBa_2Cu_3O_7$ ("YBCO" or "Y123") and $Bi_2Sr_2CaCu_2O_8$ ("BSCCO" or "Bi2212"). YBCO's critical temperature is 93 K, and BSCCO's critical temperature is 105 K at atmospheric pressure [8, 17]. The highest $T_c$ known at atmospheric pressure is in the compound $HgBa_2Ca_2Cu_3O_3$, with the critical temperature of 135 K [18] (Fig. 2.3).

$LaBa_2CuO_{4-x}$ was the first High Temperature Superconductor, discovered in 1986 by Bednorz and Muller [19] who shared the 1987 Nobel Prize in physics.

BSCCO was the first High Temperature Superconducting material used to produce HTS wires. In 1987, C. Chu et al. discovered the superconducting material YBCO, which has a $T_c$ higher than the boiling point of liquid nitrogen (77 K) [8, 20]. The development of YBCO significantly improved the performance of High Temperature Superconductors, and its manufacturing technology has already been industrialised [21].

In 2001, magnesium diboride was discovered by Nagamastsu [22]. $MgB_2$ has a critical temperature up to 39 K, which is the highest $T_c$ determined by a non copper-oxide superconductor [22, 23]. The low cost and abundance of magnesium and boron has generated new interest in $MgB_2$ for power applications [23]. In 2008, the

| **Table 2.1** Distinction between LTS and HTS [11, 53] | Low Temperature Superconductor (LTS) | High Temperature Superconductor (HTS) |
|---|---|---|
| | $T_c$ is lower than 30 K | $T_c$ is higher than 30 K |

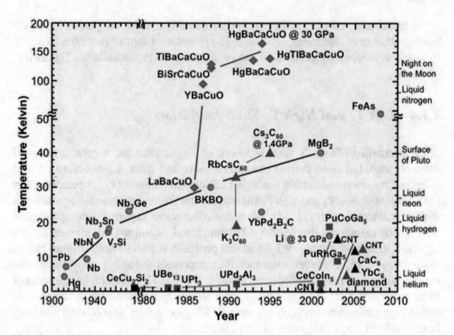

**Fig. 2.3** Timeline for the discovery of superconducting materials and the development of critical temperature [70]

**Fig. 2.4** Magnetic field vs temperature curves for five main commercial superconductors [71]

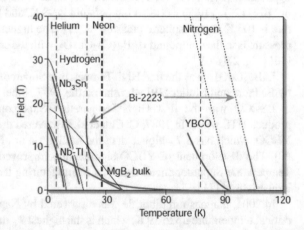

phenomenon was found that a class of iron based materials also have superconducting characteristics. These new materials are a new category of superconducting materials [23, 24].

Figure 2.4 presents "magnetic field vs temperature curves" for five main practical superconductors, and Table 2.2 shows the basic parameters for these superconductors.

**Table 2.2** Basic parameters for five practical superconductors [1, 2, 8, 71]

| Materials | $T_c$ (K) | $\mu_0 H_c 2$ (T) | $\mu_0 H_c 1$ (T) | Coherence Length (nm) | Penetration Depth (nm) | Critical Current ($A/m^{-2}$) |
|---|---|---|---|---|---|---|
| NbTi | 9 | 12 (4 K) | 10.5 (4 K) | 4 | 240 | ~4 × $10^5$ |
| Nb$_3$Sn | 18 | 27 (4 K) | 24 (4 K) | 3 | 65 | ~$10^6$ |
| MgB$_2$ | 39 | 15 (4 K) | 8 (4 K) | 6.5 | 140 | ~$10^6$ |
| YBCO | 92 | >100T (4 K) | >5T (77 K) | 1.5 | 150 | ~$10^7$ |
| Bi2223 | 108 | >100T (4 K) | >0.2T (77 K) | 1.5 | 150 | ~$10^6$ |

### 2.1.5 First Generation (1G) and Second Generation (2G) HTSs

BSCCO is also known as First Generation High Temperature Superconducting material (1G HTS) [25]. As shown in Fig. 2.5a, 1G HTS uses the structure of multi-filamentary composite [26]. From Fig. 2.5b, it can be seen that the cross-section of 1G HTS filaments is in approximate elliptical shape. Generally, there are two types of BSCCO which are processed into wires (via the powder-in-tube process). They are: Bi$_2$Sr$_2$CaCu$_2$O$_8$ (BSCCO-2212) and Bi$_2$Sr$_2$Ca$_2$Cu$_3$O$_{10}$ (BSCCO-2223) [8, 27]. 1G HTS has been the basis of earliest demonstrations of electrical power devices [26].

The coated conductor method is one of the most promising methods for producing YBCO superconducting tape, which involves deposition of YBCO on flexible metal tapes coated by multiple buffering layers [28]. YBCO tapes are defined as the Second Generation High Temperature Superconducting material (2G HTS) [1, 8, 29]. As shown in Fig. 2.5a, the biaxial texture of 2G HTS is the key to achieve high current-carrying capacity [26]. These years, YBCO 2G HTS tapes are manufactured by many companies all over the world, such as Superpower, American Superconductor, European Advanced Superconductors, Fujikura and Nexans Superconductors [8, 21, 28, 30, 31].

### 2.1.6 Type I and Type II Superconductors

As described in the previous section, superconductors enter the Meissner State when an external magnetic field is applied. But the Meissner effect will breaks down when the external field increases beyond a certain value. Type I superconductors will return to a normal state, while Type II superconductors enter a mixed state [5, 32]. Therefore, Type I is defined as the superconductors that only have the superconducting state and the normal state. However, Type II superconductors have three states: the superconducting state, the mixed state and the normal state [5, 33]. Figure 2.6 presents the difference between Type I superconductors and Type II superconductors.

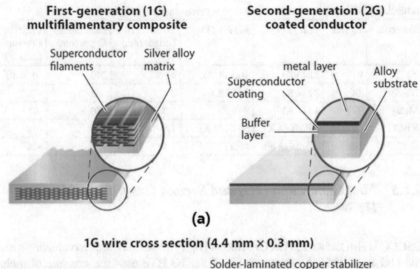

**First-generation (1G)**
**multifilamentary composite**

**Second-generation (2G)**
**coated conductor**

Superconductor    Silver alloy
filaments         matrix

metal layer        Alloy
Superconductor                substrate
coating

Buffer
layer

**(a)**

**1G wire cross section (4.4 mm × 0.3 mm)**

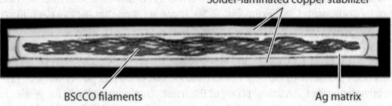

Solder-laminated copper stabilizer

BSCCO filaments                                        Ag matrix

**2G wire cross section (4.4 mm × 0.2 mm)**

1 µm YBCO HTS layer

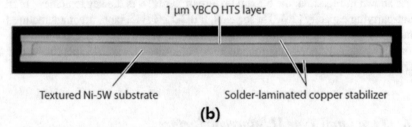

Textured Ni-5W substrate              Solder-laminated copper stabilizer

**(b)**

**Fig. 2.5  a** Schematic and **b** Cross-section of First Generation (1G) HTS and Second Generation (2G) HTS wire [26]

In Fig. 2.6a, Type I superconductors part, $B_c$ refers to the critical field. The superconductor abruptly reverts from the Meissner State to the normal state once the external applied field exceeds $B_c$, and the magnetic flux simultaneously penetrates the entire sample. The magnetization of the sample also drops to zero, thus the value of the internal magnetic field becomes equal to the external field [5, 34]. These Type I superconducting materials are mostly pure metals and metalloids [5].

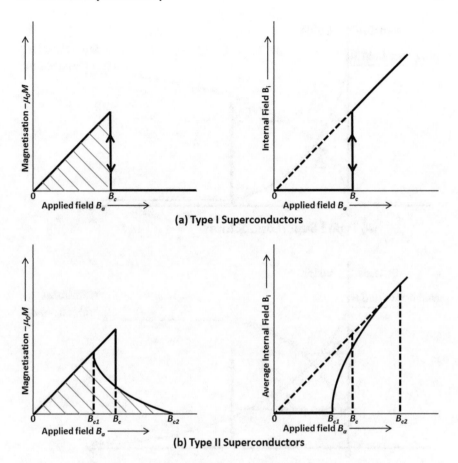

**Fig. 2.6** Magnetization and internal field of type I and type II superconductors

Type II superconductors have two critical fields: $B_{c1}$ refers to the lower critical field, and $B_{c2}$ refers to the upper critical field [5, 35]. Once the external field exceeds the lower critical field, the magnetic field starts to partially penetrate the superconductor in the form of quantised flux. The mixed state is defined as the state involving partial penetration of magnetic field, where the magnetization of the superconductor decreases gradually with an increasing external magnetic field. The magnetisation reaches zero when the external applied field reaches the upper critical field $B_{c2}$. At critical field $B_{c2}$, the external field equals the internal field [5, 35].

The Ginzburg-Landau parameter $\kappa$ can be calculated using the London penetration depth $\lambda$ and coherence length $\xi$ (shown in Fig. 2.7) [5, 36]:

$$\kappa = \frac{\lambda}{\xi} \tag{2.24}$$

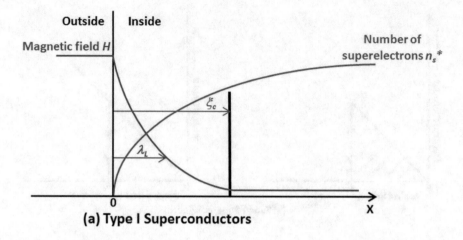

**(a) Type I Superconductors**

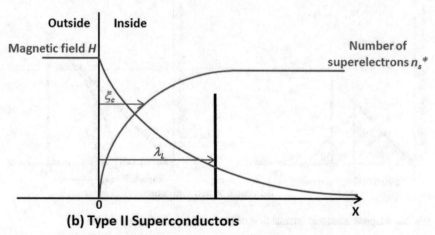

**(b) Type II Superconductors**

**Fig. 2.7** Type I and Type II superconductors: relation between the penetration depth λ and coherence length ξ

| | Type I Superconductors | Type II Superconductors |
|---|---|---|
| **Table 2.3** Distinction between type I and type II superconductors [29, 46] | $\kappa < \frac{1}{\sqrt{2}}$ | $\kappa > \frac{1}{\sqrt{2}}$ |

For type I and II superconductors, the difference in $\kappa$ is quite dramatic. E.g. a type I superconductor like lead has $\kappa = 0.48$, while a copper based type II superconductor, such as YBCO, has $\kappa = 95$ [5, 37] (Table 2.3).

The Ginsburg-Landau theory precisely describes the phenomenon of superconducting vortices during the mixed state. Generally, the size of these vortices is very small, and there are millions of vortices in a Type II bulk [5, 38].

### 2.1.7 BCS Theory

The BCS theory was named after J. Bardeen, L. Cooper and R. Schrieffer, who were the first to put forward the microscopic theory of superconductivity describing the phenomenon at the quantum mechanical level [6, 10].

According to BCS theory, super current is transported by a pair of electrons, known as Cooper pairs [6]. The Cooper pair is more stable than a single electron, and it is more insubordinate to vibrations within the material lattice [6]. The attraction to a Cooper pair's partner maintains its stability, thus it experiences less resistance. As a result, Cooper pairs passing through the material lattice are less affected by thermal vibrations below the critical temperature [6, 39].

The BCS theory reasonably explains the physical relations of the energy gap $E$ at temperature $T$ and with the critical temperature $T_c$, and the Meissner Effect as well [6, 39]. However, BCS theory only perfectly matches the physical phenomenon for low temperature superconductors, elements, and simple alloys. BCS theory cannot explain the physical phenomenon of high temperature superconductors [10].

### 2.1.8 Flux Pinning

Most applications of type II superconductors rely on their mixed state, in which they can deliver high currents in strong magnetic fields [40]. These practical type II superconductors are called hard superconductors, which is different from the ideal type II superconductors.

When an ideal type II superconducting sample is in the mixed state, this means some flux has already penetrated into the sample, which is oriented perpendicular to the plate [10, 41]. The electrical currents flowing cause a Lorentz force acting between the current and vortices:

$$F_L = J \times B \tag{2.25}$$

As the currents are spatially fixed by the boundaries, Eq. (2.25) means that the vortices must move perpendicular to the current direction and to the magnetic field [42]. The motion of vortices across the superconductor induces an electric field:

$$E = B \times v \tag{2.26}$$

That electric field is parallel to the current, thus the sample obtains electrical resistance. The motion of vortices generates heat dissipation, which is potentially owing to the losses from two fundamental factors [10]. The first loss is correlated with the appearance of electric fields which are generated by the moving vortices. These electric fields are able to accelerate the unpaired electrons, and transmit energy from the electric field to the lattice, and therefore generate heat [43]. The second loss is associated with the spatial variation for the Cooper pair density, which increases from zero in the centre of the vortices toward their outside [43].

In an ideal type II superconductor the vortices are able to move with no restriction, thus any arbitrary small currents will cause motion of the vortices and energy dissipation [43]. Therefore, for an ideal type II superconductor in the mixed state, the transport current is equal to zero. In order to make practical use of type II superconductors, they must have the capability to carry large critical currents [44, 45]. Large critical currents induce large Lorentz forces, thus a strong pinning force is necessary to repel the Lorentz forces and prevent the movement of vortices [43]. Type II superconductors with a strong pinning force can be achieved by adding impurities and crystalline defects into the materials. These materials are defined as hard superconductors.

Hard superconductors can carry large electrical currents because the magnetic fields are pinned in the volume of the superconducting material. Once a DC current (below the critical current) is flowing through in a hard superconductor, it demonstrates the lossless attribute [46]. Nevertheless, in the AC regime, vortices must move to follow the variation of the magnetic field. The pinning force characterises an obstacle [46]. The accompanied power dissipation is a hysteresis loss for hard superconductors, and will be explained in the following sections.

## 2.2   AC Loss of Superconductivity

Alternating current (AC) losses are crucial problems for superconducting applications when they are operating under the action of AC currents and AC magnetic fields [47]. AC losses can generate thermal dissipation and negatively affect overall electric power efficiency [48, 49]. As shown in Fig. 2.8, the practical superconductors, e.g. ReBCO-coated HTS tapes, consist of different materials in addition to the superconducting material, such as metals, magnetic materials, substrates, and buffers [46]. Some of them can significantly affect the total losses based on specific operating conditions. These loss contributions can be divided into four categories, which are listed below [46]:

(1) Hysteresis loss: caused by the penetration and movement of the magnetic flux in the superconducting material.
(2) Coupling loss: generated by the currents coupling two or more superconducting filaments via the normal metal regions which separate them.
(3) Ferromagnetic loss: caused by the hysteresis cycles in magnetic material parts.
(4) Eddy-current loss: arising from currents induced by a magnetic field and circulating in the normal metal parts.

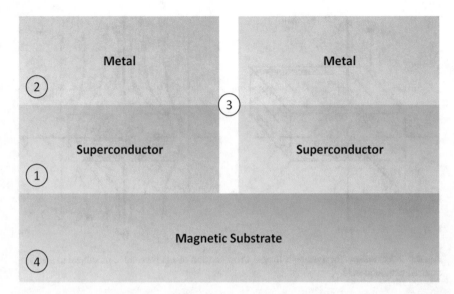

**Fig. 2.8** Schematic of the different loss contributions in practical HTSs. (1) Hysteresis loss in the superconducting parts; (2) eddy current loss in the normal metal parts; (3) coupling loss between filaments; and (4) ferromagnetic loss in the magnetic substrates

## 2.2.1   Hysteresis Loss

Hysteresis loss is generally the most important loss in the study of superconductivity. Alternating transport currents and alternating magnetic fields cause hysteresis losses in Type-II superconductors. The physical mechanisms of hysteresis losses are described in this section.

Figure 2.9 illustrates a sample of superconductor subjected to an external magnetic field. As shown in the form of flux lines, the magnetic field penetrates the superconductor [50]. The flux-line pattern and internal magnetic field change with the changing magnetic field. According to Faraday's law, the varying magnetic field inside the material induces an electric field, $E$:

$$\nabla \times E = -\frac{\partial B}{\partial t} \tag{2.27}$$

As shown in Fig. 2.9a, the electric field in the material drives "screening currents", and these "screening currents" are indicated by arrows. The direction of the currents is indicated by $+$ and $-$signs in the cross-section of the sample [50]. On the basis of Ampere's law, the screening currents determine the magnetic-field distribution in the superconductor:

$$\nabla \times B = \mu_0 J \tag{2.28}$$

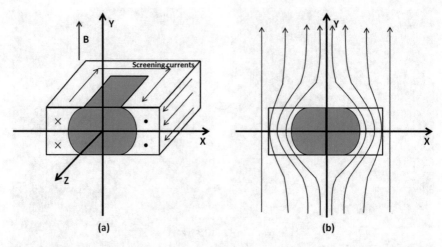

**Fig. 2.9** Mechanisms for hysteresis losses: Cross-section of a superconductor subject to a varying external magnetic field

The screening currents have power dissipation at a local power density can be calculated by $E \times J$. The energy is supplied by the power source of the magnet and delivered by the external magnetic field. Energy is required for depinning of the vortices and the movement of the flux lines, which is a dissipative process [50]. Eventually, the energy is converted to heat and must be removed by the cryogenic system.

Figure 2.9b presents the magnetic field $B$ inside and around a superconductor, assuming that the superconductor is cooled down and then subjected to a swept magnetic field with a constant changing rate. The magnetic intensity $H$ created by the external magnet is assumed to be homogeneous. $M$ is the magnetisation given by the screening currents. Therefore, the local magnetic field around the superconductor $B$ is determined by $\mu_0(H + M)$. The screening currents tend to shield the interior of the superconductor from the magnetic field variation [50]. $M$ is oriented opposite to $H$ in the centre of the superconductor, and the local magnetic field intensity is smaller than $\mu_0 H$. The lines of $M$ are closed curves from inside extending outside the sample. In the plane $y = 0$, the lines of $M$ are oriented parallel to $H$ as the symmetry, thus the magnetic field intensity is higher than $\mu_0 H$ outside the superconductor in the plane $y = 0$ [50]. Screening currents, magnetic fields and power dissipation occur only in the white region in Fig. 2.9.

Hereafter, the symbol $B$ is used to stand for the external magnetic field far away from the superconductor. Therefore, $M$ is zero, and therefore $B$ is equal to $\mu_0 H$. The screening currents provide the sample a magnetic moment $m$, which is computed from the screening current distribution. The AC loss of the superconducting sample can be calculated by the integration of either the product $B \times dm$ or $m \times dB$ over a single cycle of magnetic field. This method can also be used for the calculation of magnetisation loss due to the hysteresis of a ferromagnetic material sample [50].

A transport current flowing through a superconductor generates a magnetic field around it, which is called the self-field. If, with an AC transport current, the AC self-field also penetrates the superconductor during every cycle. Even if there is no external AC magnetic field, the variation of the AC self-field within the superconductor causes a hysteresis loss [51].

## 2.2.2 Coupling Loss

An external AC magnetic field induces eddy current in a normal conductor. There is a different sort of eddy current induced in a superconductor consisting of separate filaments embedded into a normal material. Figure 2.10 presents the cross-section of a composite conductor with two superconducting filaments indicated in grey colour. If a magnetic field is oriented perpendicular to the plane, electrical fields are induced around the loop in the plane [46]. The electric fields then induce electric currents indicated by the arrows in the loop, as shown in Fig. 2.10. Inside the superconducting filaments, the currents theoretically flow with zero resistance. Resistance only occurs at the joints of the composite where electrical currents flow across the normal conducting matrix. The cross currents shown in Fig. 2.10 are called coupling currents since they couple the superconducting filaments together with normal materials into a single large system. This additional power dissipation is generated as the coupling currents flow, which is defined as the coupling current loss [50].

In order to decrease the coupling current AC loss, the coupling currents should be controlled below the critical current of the superconducting filaments. In order to reduce coupling loss, twisting the filaments and reducing the dimensions of the composite can be carried out [52].

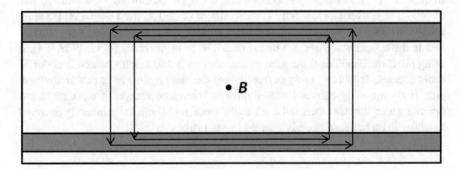

**Fig. 2.10** Coupling AC loss in composite conductor

### 2.2.3  Ferromagnetic Loss

It is common that magnetic materials appear in superconducting systems because magnetic materials can be important components of superconducting devices, or a part of superconducting wires [46]. For example, some Second Generation superconducting tapes can be manufactured using substrates based on ferromagnetic materials such as alloys of Ni and W, and this causes additional losses which must be taken into consideration when estimating the total losses in the HTS tapes. The magnetic components affect the AC loss in superconducting systems in two ways [46]:

(1)  They adjust the magnetic profile inside the superconductor, such as the iron core in superconducting motors.
(2)  They contribute additional hysteretic losses to the system, in the magnetic substrates.

Therefore, the contribution of the magnetic components on the total AC losses in a superconducting system is important.

### 2.2.4  Eddy-Current Loss

Eddy current losses are originated in terms of power dissipation by the resistive (Joule) effect. If a metal conductor is exposed to an alternating external magnetic field, there will be an eddy-current induced, and this eddy-current as well as corresponding material's electrical resistance cause power dissipation [46].

Some studies have proved that eddy-current losses are negligible at low operating frequencies, but they should be taken into consideration when working at high frequencies [53]. For example, considering a Copper Stabilized Tape with 50 μm of copper stabilizer and self-field critical current of 260 A, used with a 60 Hz sinusoidal current with peak current of critical current, the corresponding eddy-current loss in the copper stabilizer can be estimated to be in the order of 5 mW/m [54]. If using the same conditions, the total hysteresis loss is 380 mW/m, which is about 32 times greater. Therefore, eddy-current losses are not significant at power frequencies. However, eddy-current losses should be taken into account if working at low operating temperature (such as 4.2 K), as the metal resistivities significantly decrease resulting in an increase in eddy-current losses [46].

## 2.3  AC Loss Measurements

Basically, there are three common methods for AC loss measurement of superconductors: the electric method, the magnetic method, and the calorimetric method. Their advantages and drawbacks are briefly demonstrated below.

### 2.3.1   Electrical Method

If an AC current source supplies the electric energy, and some of the energy is dissipated in the superconductor due to flux creep or self-field loss [50]. Their electric energy $q_{trans}$ (in Joule) dissipated during each cycle is called the transport current loss. The electrical method can be used to measure the transport current loss by measuring the voltage in-phase with the current, which can be assumed as a resistive load [52]. A compensation coil is used to cancel the inductive signal, when the measuring superconductor has inductive quantity induce voltages which are not due to the transport current loss. Additionally, a strong alternating magnetic field can disturb the electric measurement. This thesis focuses on the electrical method for AC loss measurement, and more details on the electrical method are presented in Chaps. 7 to 9.

### 2.3.2   Magnetic Method

The AC magnetic source provides magnetic energy, and some of the energy is dissipated in the superconductor due to hysteresis or coupling current losses. The magnetic energy $q_{magn}$ (in Joule) dissipated during each cycle is defined as the magnetisation loss [50]. The variation in the magnetic moment of the superconducting sample is determined by the voltages of pickup coils around the sample [55]. However, the small AC transport current detected by the magnetic method may affect the performance of a magnetic measurement, because it adds an extra magnetic moment to the sample which is not owing to the external magnetic field.

### 2.3.3   Calorimetric Method

The calorimetric method can be carried out either by measuring the temperature increment of the superconducting sample, or by detecting the amount of nitrogen that is evaporated from the cryostat [52]. The calorimetric method has the advantage that it cannot be affected by undesirable alternating currents or magnetic field disturbances, as mentioned above. However, less information can be acquired using the calorimetric method, because the magnetisation loss cannot be distinguished from the total losses. The calorimetric method can also be affected by thermal influences from the external environment due to Ohmic dissipation in the current leads, or heat leakage from the cryostat [50].

## 2.4   Analytical Techniques for AC Loss

Compared to AC loss measurements, a theoretical estimation of AC loss is a much more efficient way to obtain a reliable reference for the approximate amount of AC loss in a superconductor. Analytical techniques also possess the advantage over AC loss measurement that they are not affect by ambient factors and experimental errors. There are two typical analytical solutions, "Norris" and "Brandts", for estimating the hysteresis AC loss of a superconductor under the action of AC transport current and AC external magnetic field. Both are widely used in the superconductivity research community.

### 2.4.1   Norris Analytical Solutions

Norris analytical solutions are general methods to estimate the hysteresis AC loss in a Type-II superconductor under the action of AC transport current [51]. There are two scenarios for which Norris analytical solutions are frequently used, Norris Ellipse and Norris Strips, as they are the most general geometries of superconducting tapes.

For Type-II elliptical superconducting tapes, if the critical current is $I_c$, and the applied current is $I_a$, the Norris Ellipse analytical solution is:

$$Q_{(Joule/cycle/m)} = \frac{I_c^2 \mu_0}{\pi}\left[\left(1 - \frac{I_a}{I_c}\right)\ln\left(1 - \frac{I_a}{I_c}\right) + \left(2 - \frac{I_a}{I_c}\right)\frac{I_a}{2I_c}\right] \quad (2.29)$$

For Type-II rectangle strip superconducting tapes, the Norris Strip analytical solution is:

$$Q_{(Joule/cycle/m)} = \frac{I_c^2 \mu_0}{\pi}\left[\left(1 - \frac{I_a}{I_c}\right)\ln\left(1 - \frac{I_a}{I_c}\right) + \left(1 + \frac{I_a}{I_c}\right)\ln\left(1 + \frac{I_a}{I_c}\right) - \left(\frac{I_a}{I_c}\right)^2\right]$$
$$(2.30)$$

### 2.4.2   Brandt Analytical Solutions

Brandt analytical solutions are general approaches to calculate hysteresis AC losses for a Type–II superconducting strip subject to a perpendicular external magnetic field [56]. For a superconducting strip of width $2a$ and thickness $d$ ($2a \gg d$), with critical magnetic field $H_c$, exposed to a perpendicular magnetic field $H_a$, the magnetization loss is:

$$P_{(Watt/m)} = 4\pi \mu_0 a^2 H_a H_c f \left[ \frac{2H_c}{H_a} \ln \cosh\left(\frac{H_c}{H_a}\right) - \tanh\left(\frac{H_c}{H_a}\right) \right] \qquad (2.31)$$

## 2.5 Historical Development of Lorentz Force Electrical Impedance Tomography

Lorentz Force Electrical Impedance Tomography (LFEIT), also known as Hall Effect Imaging (HEI) or Magneto-Acousto-Electric Tomography (MAET), is one of the most promising hybrid methods with burgeoning potential for biological imaging, particularly in cancer detection [57–60]. Simple ultrasonic imaging technology has difficulty in distinguishing between soft tissues as acoustic impedance varies by less than 10% between muscle and blood. LFEIT shows the powerful capability to provide information about the pathological and physiological condition of tissue, because electrical impedance varies widely among soft tissue types and pathological states [57, 61–63]. In addition, tissues under conditions of haemorrhage or ischemia exhibit huge difference in electrical properties because most body fluid and blood have fairly different permittivities and conductivities compared to other soft tissues [64, 65]. The detailed data of electrical resistivity of human tissues (at 20 to 200 kHz) as well as the acoustic properties of human tissues are given in Tables 2.4 and 2.5, respectively [64, 66].

In the 1990s, Hall Effect Imaging (HEI) was developed by Wen et al. [58]. HEI uses the Lorentz Force based coupling mechanism with ultrasound to image the electrical properties of biological tissues. The Hall Effect explains the phenomenon of charge separation in a conductive object subjected to a magnetic field. Figure 2.11 demonstrates an experiment of Hall Effect Imaging [58, 64]. HEI technology detects the Hall voltages using surface electrodes on the tested specimen, where these voltages

**Table 2.4** Electrical resistivity of human tissues (at 20 to 200 kHz) [72]

| Human Tissue | Electrical Resistivity ($\Omega \cdot m$) |
|---|---|
| Blood | 1.5 |
| Blood plasma | 0.66 |
| Bone | 166 |
| Cardiac muscle (Longitudinal) | 1.60–5.65 |
| Cardiac muscle (Transverse) | 4.25–51.81 |
| Fat | 21–28 |
| Liver | 3.6–5.5 |
| Lung (breath in) | 7.27 |
| Lung (breath out) | 23.63 |
| Skeletal muscle (Longitudinal) | 1.25–1.5 |
| Skeletal muscle (Transverse) | 18–23 |

**Table 2.5** The acoustic properties of human tissues [72]

| Human Tissue | Mass Density (g/m) | Sound Speed in Tissue (m/s) | Acoustic Characteristic Impedance | Frequency (MHz) |
|---|---|---|---|---|
| Blood | 1.055 | 1570 | 1.655 | 1.0 |
| Bone | 1.658 | 3860 | 5.571 | 1.0 |
| Fat | 0.955 | 1476 | 1.410 | 1.0 |
| Liver | 1.050 | 1570 | 1.648 | 1.0 |
| Muscle | 1.074 | 1568 | 1.685 | 1.0 |
| Soft tissue | 1.016 | 1500 | 1.525 | 1.0 |

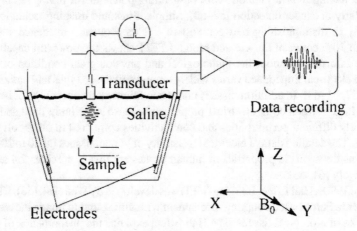

**Fig. 2.11** Experiment of Hall Effect imaging [58]

are induced using ultrasound to cause localised mechanical vibrations in a conductive tissue specimen located in a static magnetic field. For biological tissue samples, the Hall voltages are induced by "the Lorentz force induced separation of conductive ions in intra- and extra-cellular space" [64, 67]. With the ultrasound packet propagating through the specimen, the conductivity data of the specimen along the ultrasound beam can be encoded in the time course of measuring the Hall voltages [58, 64, 67]. HEI technology is able to realise high spatial resolution images with respect to the conductivity distribution within biological tissue specimens. The spatial imaging of HEI is very close to ultrasound imaging, as it is mainly determined by the bandwidth and central frequency of the ultrasound packets generated [58, 59, 64, 67].

The method of Magneto-acoustic tomography with magnetic induction (MAT-MI) was proposed by Bin He et al. This breakthrough solved the shielding effect problem which existed in other hybrid bio-conductivity imaging techniques such as HEI (Xu and He 2005) [60]. Figure 2.12 presents the schematic of Magneto-acoustic tomography with magnetic induction. Unlike HEI, MAT-MI uses the Lorentz force to induce an eddy current to produce ultrasound vibrations which can be detected using

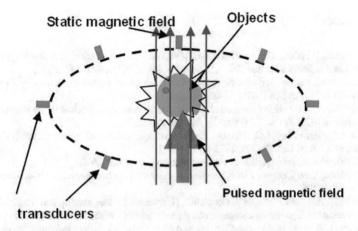

**Fig. 2.12** Schematic of Magneto-acoustic tomography with magnetic induction [60]

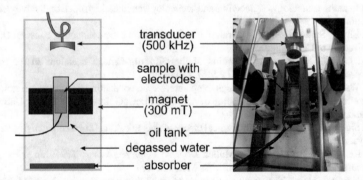

**Fig. 2.13** Experimental design of Lorentz Force Electrical Impedance Tomography [59]

ultrasound transducers (receiving mode) placed around the specimen. The recorded ultrasound signals are then used to reconstruct the conductivity distribution of the biological sample [60, 68].

In 2013, a small scaled experimental Lorentz Force Electrical Impedance Tomography (LFEIT) was developed by Grasland-Mongrain et al. [59]. As shown in Fig. 2.13, this design collected electric current from electrodes placed around the specimens. Two specimens were chosen: a gelatin phantom and a beef sample, which were successively fixed into a 0.3 T magnetic field $B_0$ and sonicated with an ultrasonic transducer emitting 500 kHz bursts [59, 69].

# References

1. P.A. Abetti, P. Haldar, One hundred years of superconductivity: science, technology, products, profits and industry structure. Int. J. Technol. Manage. **48**(4), 423–447 (2009)
2. D. van Delft, History and significance of the discovery of superconductivity by Kamerlingh Onnes in 1911. Physica C **479**, 30–35 (2012)
3. M.N. Wilson, 100 years of superconductivity and 50 years of superconducting magnets, Appl. Supercond. IEEE Trans. **22**(3) (2012), Art. no. 3800212
4. D. Domínguez, E. Jagla, C. Balseiro, Phenomenological theory of the paramagnetic Meissner effect. Phys. Rev. Lett. **72**(17), 2773–2776 (1994)
5. A.C. Rose-Innes, *Introduction to superconductivity*. Elsevier (2012)
6. J. Bardeen, L.N. Cooper, J.R. Schrieffer, Theory of superconductivity. Phys. Rev. **108**(5), 1175–1204 (1957)
7. C. Jooss, J. Albrecht, H. Kuhn, S. Leonhardt, H. Kronmüller, Magneto-optical studies of current distributions in high-Tc superconductors. Rep. Prog. Phys. **65**(5), 651–788 (2002)
8. Z. Zhong, A study of Critical Currents and Quench in 2G Superconductors under varying magnetic fields, Ph.D. Thesis, University of Cambridge (2015)
9. T.O. University, The London equations, 18 August, 2015. http://www.open.edu/openlearn/sci ence-maths-technology/engineering-and-technology/engineering/superconductivity/content-section-3.3
10. M. Zhang, Study of second generation high temperature superconducting coils, Ph.D. Thesis, University of Cambridge (2013)
11. L.P. Gorkov, Microscopic derivation of the Ginzburg-Landau equations in the theory of superconductivity. Sov. Phys. JETP **9**(6), 1364–1367 (1959)
12. W. Wang, An investigation into high temperature superconducting flux pump technology with the circular type magnetic flux pump devices and YBaCuO films, Ph..D Thesis, University of Cambridge (2014)
13. M. Sigrist, K. Ueda, Phenomenological theory of unconventional superconductivity. Rev. Mod. Phys. **63**(2), 239–311 (1991)
14. E. Kaldis, E. Liarokapis, K.A. Müller, *High-Tc superconductivity 1996: ten years after the discovery*: Springer Science & Business Media (2012)
15. G. Hardy, J. Hulm, The superconductivity of some transition metal compounds. Phys. Rev. **93**(5), 1004–1016 (1954)
16. D. Larbalestier, The road to conductors of high temperature superconductors: 10 years do make a difference! Appl. Supercond. IEEE Trans. **7**(2), 90–97 (1997)
17. W. Gao, Z. Li, N. Sammes, *Superconductivity and Superconducting Materials*
18. R. Wesche, *High-temperature superconductors: materials, properties, and applications.* Springer Science & Business Media (2013)
19. J.G. Bednorz, K.A. Müller, Possible highT c superconductivity in the Ba–La–Cu–O system. Zeitschrift für Physik B Condensed Matter **64**(2), 189–193 (1986)
20. S. Martin, A. Fiory, R. Fleming, L. Schneemeyer, J. Waszczak, Temperature dependence of the resistivity tensor in superconducting Bi 2 Sr 2.2 Ca 0.8 Cu 2 O 8 crystals. Phys. Rev. Lett. **60**(21), 2194–2197 (1988)
21. D.M. Gann, Construction as a manufacturing process? Similarities and differences between industrialized housing and car production in Japan. Constr. Manag. Econ. **14**(5), 437–450 (1996)
22. S. Jin, H. Mavoori, C. Bower, R. Van Dover, High critical currents in iron-clad superconducting MgB2 wires. Nature **411**(6837), 563–565 (2001)
23. S. BudKo, G. Lapertot, C. Petrovic, C. Cunningham, N. Anderson, P. Canfield, Boron isotope effect in superconducting MgB 2. Phys. Rev. Lett. **86**(9), 1877–1880 (2001)
24. D. Larbalestier, L. Cooley, M. Rikel, A. Polyanskii, J. Jiang, S. Patnaik, X. Cai, D. Feldmann, A. Gurevich, A. Squitieri, Strongly linked current flow in polycrystalline forms of the superconductor MgB2. Nature **410**(6825), 186–189 (2001)

25. M.W. Rupich, X. Li, C. Thieme, S. Sathyamurthy, S. Fleshler, D. Tucker, E. Thompson, J. Schreiber, J. Lynch, and D. Buczek, "Advances in second generation high temperature superconducting wire manufacturing and R&D at American superconductor corporation, Supercond. Sci. Technol. **23**(1) (2010), Art. no. 014015

26. A. Malozemoff, Second-generation high-temperature superconductor wires for the electric power grid. Annu. Rev. Mater. Res. **42**, 373–397 (2012)

27. N. Basturk, Electrical and structural properties of Ex-situ annealed superconducting Bi2Sr2CaCu2O8 thin films obtained by coevaporation of components. Turk. J. Phys **29**, 115–118 (2005)

28. M. Rupich, D. Verebelyi, W. Zhang, T. Kodenkandath, X. Li, Metalorganic deposition of YBCO films for second-generation high-temperature superconductor wires. MRS Bull. **29**(08), 572–578 (2004)

29. K. Kim, M. Paranthaman, D.P. Norton, T. Aytug, C. Cantoni, A.A. Gapud, A. Goyal, D.K. Christen, A perspective on conducting oxide buffers for Cu-based YBCO-coated conductors. Supercond. Sci. Technol. **19**(4), 23–29 (2006)

30. Y. Xie, A. Knoll, Y. Chen, Y. Li, X. Xiong, Y. Qiao, P. Hou, J. Reeves, T. Salagaj, K. Lenseth, Progress in scale-up of second-generation high-temperature superconductors at SuperPower Inc. Physica C **426**, 849–857 (2005)

31. T. Honjo, H. Fuji, Y. Nakamura, T. Izumi, T. Araki, Y. Yamada, I. Hirabayashi, Y. Shiohara, Y. Iijima, K. Takeda, A metal substrate of non-magnetism or weak magnetism and high strength, are deposited with an intermediate layer having a high alignment, depressing the diffusion and reaction of elements, Google Patents (2003)

32. J. Bardeen, M. Stephen, Theory of the motion of vortices in superconductors. Phys. Rev. **140**(4A), 1197–1207 (1965)

33. G. Blatter, M. FeigelMan, V. Geshkenbein, A. Larkin, V.M. Vinokur, Vortices in high-temperature superconductors. Rev. Mod. Phys. **66**(4), 1125–1388 (1994)

34. W. Buckel, R. Kleiner, *Outlook of the 1st Edition (1972)*: Wiley Online Library (1991)

35. A. Gurevich, Enhancement of the upper critical field by nonmagnetic impurities in dirty two-gap superconductors, Phys. Rev. B **67**(18) (2003), Art. no. 184515

36. G. Eilenberger, Determination of $\kappa$ 1 (T) and $\kappa$ 2 (T) for Type-II Superconductors with Arbitrary Impurity Concentration. Phys. Rev. **153**(2), 584–598 (1967)

37. H. Eisaki, N. Kaneko, D. Feng, A. Damascelli, P. Mang, K. Shen, Z.-X. Shen, M. Greven, Effect of chemical inhomogeneity in bismuth-based copper oxide superconductors, Phys. Rev. B **69**(6) (2004), Art. no. 064512

38. C. Antoine, *Materials and surface aspects in the development of SRF Niobium cavities* (2012)

39. H. Suhl, B. Matthias, L. Walker, Bardeen-Cooper-Schrieffer theory of superconductivity in the case of overlapping bands. Phys. Rev. Lett. **3**(12), 552–554 (1959)

40. R.M. Scanlan, A.P. Malozemoff, D.C. Larbalestier, Superconducting materials for large scale applications. Proc. IEEE **92**(10), 1639–1654 (2004)

41. P.W. Anderson, Y. Kim, Hard superconductivity: theory of the motion of Abrikosov flux lines. Rev. Mod. Phys. **36**(1), 39–43 (1964)

42. C.J. Gorter, Lorentz Force and Flux Motion. Phys. Lett. **1**(69), 121–143 (1962)

43. W. Buckel, R. Kleiner, *Superconductivity: Fundamentals and applications*. Wiley (2004)

44. D. Larbalestier, A. Gurevich, D.M. Feldmann, A. Polyanskii, High-Tc superconducting materials for electric power applications, *Materials For Sustainable Energy: A Collection of Peer-Reviewed Research and Review Articles from Nature Publishing Group*, pp. 311–320: World Scientific (2011)

45. R. Garwin, J. Matisoo, Superconducting lines for the transmission of large amounts of electrical power over great distances. Proc. IEEE **55**(4), 538–548 (1967)

46. F. Grilli, E. Pardo, A. Stenvall, D.N. Nguyen, W. Yuan, F. Gömöry, Computation of losses in HTS under the action of varying magnetic fields and currents. IEEE Trans. Appl. Supercond. **24**(1), 78–110 (2014)

47. B. Shen, J. Li, J. Geng, L. Fu, X. Zhang, H. Zhang, C. Li, F. Grilli, and T. A. Coombs, Investigation of AC losses in horizontally parallel HTS tapes, Supercond. Sci. Technol. **30**(7) (2017), Art. no. 075006

48. H. Yu, G. Zhang, L. Jing, Q. Liu, W. Yuan, Z. Liu, and X. Feng, Wireless power transfer with HTS transmitting and relaying coils, IEEE Trans. Appl. Supercond. **25**(3) (2015)

49. S. Lee, S. Byun, W. Kim, K. Choi, H. Lee, AC loss analysis of a HTS coil with parallel superconducting tapes of unbalanced current distribution. Phys. C, Supercond. **463**, 1271–1275 (2007)

50. M. Oomen, AC Loss in superconducting Tapes and Cables, Ph.D. Thesis, University of Twente (2000)

51. W. Norris, Calculation of hysteresis losses in hard superconductors carrying ac: isolated conductors and edges of thin sheets. J. Phys. D Appl. Phys. **3**(4), 489–507 (1970)

52. Y. Chen, AC Loss Analysis and Measurement of Second Generation High Temperature Superconductor Racetrack Coils, Ph.D. Thesis, University of Cambridge (2014)

53. B. Shen, J. Li, J. Geng, L. Fu, X. Zhang, C. Li, H. Zhang, Q. Dong, J. Ma, T.A. Coombs, Investigation and comparison of AC losses on Stabilizer-free and Copper Stabilizer HTS tapes. Phys. C, Supercond. **541**, 40–44 (2017)

54. R. Duckworth, M. Gouge, J. Lue, C. Thieme, D. Verebelyi, Substrate and stabilization effects on the transport AC losses in YBCO coated conductors. IEEE Trans. Appl. Supercond. **15**(2), 1583–1586 (2005)

55. J. Šouc, E. Pardo, M. Vojenčiak, F. Gömöry, Theoretical and experimental study of AC loss in high temperature superconductor single pancake coils, Supercond. Sci. Technol. **22**(1) (2008), Art. no. 015006

56. E.H. Brandt, M. Indenbom, Type-II-superconductor strip with current in a perpendicular magnetic field. Physical review B **48**(17), 893–906 (1993)

57. N. Polydorides, *In-Vivo Imaging With Lorentz Force Electrical Impedance Tomography* (University of Edinburgh, UK, 2014)

58. H. Wen, J. Shah, R.S. Balaban, Hall effect imaging. IEEE Trans. Biomed. Eng. **45**(1), 119–124 (1998)

59. P. Grasland-Mongrain, J. Mari, J. Chapelon, C. Lafon, Lorentz force electrical impedance tomography. IRBM **34**(4), 357–360 (2013)

60. Y. Xu, B. He, Magnetoacoustic tomography with magnetic induction (MAT-MI), Phys. Med. Biol. **50**(21) (2005)

61. Y. Zou, Z. Guo, A review of electrical impedance techniques for breast cancer detection. Med. Eng. Phys. **25**(2), 79–90 (2003)

62. C. Gabriel, S. Gabriel, E. Corthout, The dielectric properties of biological tissues: I. Literature survey, Phys. Med. Biol. **41**(11), 2231 (1996)

63. D. Holder, *Electrical impedance tomography: methods, history and applications.* CRC Press (2004)

64. X. Li, *Magnetoacoustic Tomography with Magnetic Induction for Electrical Conductivity Imaging of Biological Tissue.* University of Minnesota ()2010

65. H. Schwan, K. Foster, RF-field interactions with biological systems: electrical properties and biophysical mechanisms. Proc. IEEE **68**(1), 104–113 (1980)

66. W.T. Joines, Y. Zhang, C. Li, R.L. Jirtle, The measured electrical properties of normal and malignant human tissues from 50 to 900 MHz. Med. Phys. **21**(4), 547–550 (1994)

67. A. Montalibet, J. Jossinet, A. Matias, D. Cathignol, Electric current generated by ultrasonically induced Lorentz force in biological media. Med. Biol. Eng. Compu. **39**(1), 15–20 (2001)

68. L. Mariappan, B. He, Magnetoacoustic tomography with magnetic induction: Bioimepedance reconstruction through vector source imaging. Medical Imaging IEEE Trans. **32**(3), 619–627 (2013)

69. H. Ammari, P. Grasland-Mongrain, P. Millien, L. Seppecher, J.-K. Seo, A mathematical and numerical framework for ultrasonically-induced Lorentz force electrical impedance tomography. Journal de Mathematiques Pures et Appliquées **103**(6), 1390–1409 (2015)

70. Wikipedia. https://en.wikipedia.org/wiki/Superconductivity, 18 August, 2015

71. D. Larbalestier, A. Gurevich, D.M. Feldmann, A. Polyanskii, High-Tc superconducting materials for electric power applications. Nature **414**(6861), 368–377 (2001)
72. L. Li, Simulation for the Source Distribution of Mageto-acoustic Effect Based on ANSYS, Master Thesis, Peking Union Medical College (2011)

# Chapter 3
# Numerical Modelling and Theoretical Analysis

## 3.1 High $T_c$ Superconductors (HTS) Model and Simulation

Several models have been established for calculating the field distributions and current density and for High Temperature Superconductors (HTS), which are described in detail in the following sections.

### 3.1.1 Critical State Model

The critical state model was developed empirically based on the data from macroscopic superconductivity experiments. It offers a convenient approach to mathematically explain the superconductivity phenomena [1]. The critical state model, defined as the simplest and most widely used superconductivity model, assumes that the outer layer of the superconductivity material entirely enters the critical state when low magnetic fields or currents are applied, whilst the interior region is still in the virgin (normal) state [1, 2]. Using the critical state model is a straightforward method to understanding engineering applications of superconductivity. There are two critical state models commonly used, namely, the Bean model proposed by Bean [3], and the Kim model proposed by Kim [4], which are briefly introduced in the following sections.

B. Shen, *Study of Second Generation High Temperature Superconductors: Electromagnetic Characteristics and AC Loss Analysis*, Springer Theses,
https://doi.org/10.1007/978-3-030-58058-2_3

### 3.1.2 The Bean Model

The Bean Model is one of the simplest of the critical state models, in which two crucial assumptions are [3, 5]:

(a)  The critical current $J_c$ is independent of the magnetic field.
(b)  A critical current $\pm J_c$ flows wherever the superconducting material enters the critical state, while there is no current when the superconducting material is in its virgin (normal) state.

The Bean's approximation predicts that shielding or trapped current is only able to be the values 0 or $\pm J_c$. As shown in Fig. 3.1, the Bean model can be illustrated by a superconducting slab with y–z plane of infinite length which is exposed to an applied magnetic field in the z direction. Induced currents will flow around the surface of the geometry when the slab is in the presence of an applied magnetic field in the z direction. Because the assumption is made that the superconductor is infinitely long in the y–z plane, the surface current in the x direction is negligible compared to the current in the y direction. Therefore, the currents are assumed to flow only in the y direction (two-way) [3, 5].

When an applied magnetic field $H_a$ increases monotonically, screening currents are induced at the surface of the slab and gradually penetrate into the slab interior. This produces a magnetic field in the opposite direction to the field applied. As shown in Fig. 3.2a, the deep interior of the slab is shielded from the external field. The slope is the derivative with respect to the magnetic field inside the slab, and is equal to the critical current:

**Fig. 3.1** Superconducting slab with infinite length in the y–z plane which is in the presence of an applied magnetic field in the z direction

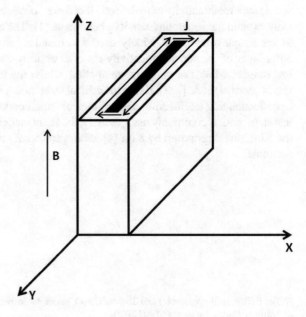

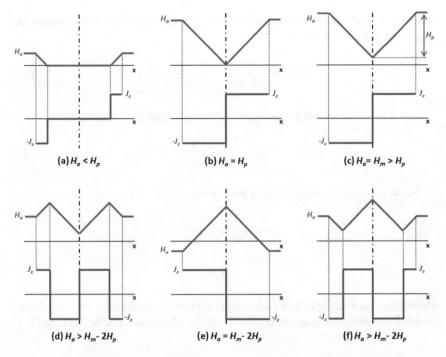

**Fig. 3.2** Magnetic field and current profiles for the superconducting slab presented in Fig. 3.1

$$\pm J_c = \frac{\partial H_y}{\partial x} \tag{3.1}$$

As shown in Fig. 3.2b, at the moment $H_a$ has increased to the penetration field $H_p$, the applied field fully penetrates the superconducting slab and a current density $J_c$ flows in each half. When the applied field is greater than $H_p$, there is an upward shifting of the field magnitude in the slab because the density of the screening current cannot exceed $J_c$, and the field gap between the edge and centre of the slab retains the same value of $H_p$ (Fig. 3.2c). Figure 3.2d illustrates the process of the decreasing field, when currents of opposite sign are induced at the outer layer of the slab due to Lenz's Law. The magnetic field and current density profiles are completely reversed when the applied field decreases by $2 \times H_p$ of its peak value (Fig. 3.2e). As shown in Fig. 3.2f, reversed current flows at the edges of the outer region of the slab if the applied magnetic field begins to increase again [3, 5, 6].

### 3.1.3   The Kim Model

In 1962, Kim extended the Bean model by incorporating the temperature and magnetic field, both of which affect the critical current. In the Kim model, the critical current $J_c$ of the superconducting material is no longer a constant at all the time, but becomes a variable [4, 7]:

$$J_c(B) = \frac{\alpha(T)}{B_0 + B} = J_{co}(T)\frac{1}{1 + \frac{B}{B_0}} \tag{3.2}$$

$B_0$ is a constant related to the superconducting material:

$$J_{co}(T) = \frac{\alpha(T)}{B_0} \tag{3.3}$$

The temperature dependence $\alpha$ is defined as:

$$\alpha = \frac{1}{d}(a - bT) \tag{3.4}$$

$$\frac{a}{b} \le T_c \tag{3.5}$$

where d is a constant dependent on the physical micro structure of the superconductor. Detailed comparisons of the Bean model and the Kim model can be found in [4, 7].

### 3.1.4   E-J Power Law

Maxwell's equations are always valid, but the superconducting material properties should be modified properly to carry out a physical model. The critical state model has shown sufficient accuracy for use with Low Temperature Superconductors [8]. However, for High Temperature Superconducting materials the critical current density $J_c$ is an ill-defined quantity. Based on experimental measurements, scientists have determined the E-J constitutive power law for Type II superconductors. Now the E-J power law has become the prevailing way to describe the electrical properties of Type II superconductors, defined as [8, 9]:

$$E = E_0\left(\frac{J}{J_c}\right)^n \tag{3.6}$$

where $E_0$ is set to be $10^{-4}$ Vm$^{-1}$. $J_c$ is the critical current density determined at the moment the electric field reaches $E_0$. Based on experimental results from material properties and microstructures, $n$ is a measured parameter used to fit the E-J power equation [10].

Figure 3.3 illustrates the relationship between $E/E_0$ and $J/J_0$ with different values of $n$. The pure ohmic case uses $n = 1$, and the critical state model corresponds to $n = \infty$. For Type II superconductors, the value of $n$ usually lies in the range from 15

**Fig. 3.3** *E-J* power law with different *n* factors [6]

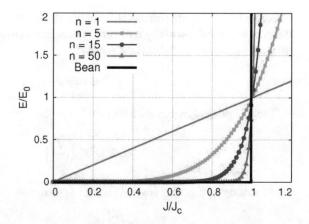

to 30. For NbTi, *n* is from 40–80. A typical value for YBCO is $n = 25$, which was selected for most modelling and design in this thesis [6].

### 3.1.5 General **H**-Formulation

Maxwell's equations are widely used for conventional electromagnetic problems to calculate the current and magnetic field distribution. Maxwell's equations are still valid for solving superconducting problems, but additional equations for describing material properties are necessary. The **H**-formulation is one of the methods used to realise superconducting modelling, which is to calculate the solution of the induced current and magnetic field distribution under the framework of the *E-J* power law. This section uses Maxwell's equations for a conventional electromagnetic problem, coupled with an *E-J* power law to extend this framework to a superconducting formulation [10, 11].

Ampere's Law explains the relationship between current and field:

$$\nabla \times \boldsymbol{H} = \boldsymbol{J} \tag{3.7}$$

where **H** is the magnetic field and **J** is the current density. Faraday s Law is written as:

$$\nabla \times \boldsymbol{E} = -\frac{\partial \boldsymbol{B}}{\partial t} \tag{3.8}$$

By constitutive law:

$$\boldsymbol{B} = \mu_0 \mu_r \boldsymbol{H} \tag{3.9}$$

where $\mu_0$ is the vacuum permeability, and $\mu_r$ is the relative permeability of the material. The electrical resistivity $\rho$ of certain materials is used to obtain the relationship between electric field and current density:

$$E = \rho J \tag{3.10}$$

The $E$-$J$ power law has been derived before as:

$$E = E_0 \left( \frac{J}{J_c} \right)^n \tag{3.11}$$

According to Eqs. (3.7) to (3.11), after adjustment:

$$\frac{\partial(\mu_0 \mu_r H)}{\partial t} + \nabla \times (\rho \nabla \times H) = 0 \tag{3.12}$$

Equation (3.12) is a single equation with a single known value of $H$, which is known as the general $H$-formulation.

### 3.1.6  Two-Dimensional (2D) H-Formulation Models

As shown in Fig. 3.4, a plane on the paper (x and y direction) is perpendicular to the space with infinite length in the z direction. A uniform magnetic field entirely in the

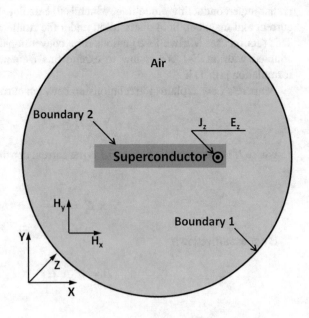

**Fig. 3.4** Boundaries and subdomains of Two-dimensional (2D) $H$-formulation Models. The electric field $E_z$ and the superconducting current density $J_z$ are in the z-direction, and the magnetic field $H_x$ and $H_y$ are in the x-y plane

y direction ($H_y$) is applied to a superconducting long rod (rectangular cross section) in the z direction [11].

The 2D Cartesian coordinates model comprises 2 variables, which are defined as $H = [H_x, H_y]$. As shown in Fig. 3.4, in this 2D model the current, $J_z$ in the superconductor flows only in the z direction [11]. The electric field $E_z$ is also in the z direction. Thus, $E_z = \rho J_z$, where $\rho$ is the resistivity of the material. Ampere's Law for the 2D model can be expressed as:

$$J_z = \frac{\partial H_y}{\partial x} - \frac{\partial H_x}{\partial y} \tag{3.13}$$

Substituting $H = [H_x, H_y]$, the Faraday's Law for this model can be written as:

$$\begin{bmatrix} \frac{\partial E_z}{\partial y} \\ -\frac{\partial E_z}{\partial x} \end{bmatrix} = -\mu_0 \mu_r \begin{bmatrix} \frac{\partial H_x}{\partial t} \\ \frac{\partial H_y}{\partial t} \end{bmatrix} \tag{3.14}$$

Using the resistivity relationship Eq. (3.10) and the $E$-$J$ power law Eq. (3.11) to substitute for electric field $E_z$, we obtain the two equations with two unknowns $H_x$ and $H_y$. This equation is similar to Eq. (3.12):

$$\begin{bmatrix} \dfrac{\partial \left( E_0 \left( \frac{\frac{\partial H_y}{\partial x} - \frac{\partial H_x}{\partial y}}{J_c} \right)^n \right)}{\partial y} \\ -\dfrac{\partial \left( E_0 \left( \frac{\frac{\partial H_y}{\partial x} - \frac{\partial H_x}{\partial y}}{J_c} \right)^n \right)}{\partial x} \end{bmatrix} = -\mu_0 \mu_r \begin{bmatrix} \frac{\partial H_x}{\partial t} \\ \frac{\partial H_y}{\partial t} \end{bmatrix} \tag{3.15}$$

Equation (3.15) involves two Partial Differential Equations (PDEs) and two dependent variables for superconducting materials. For other normal regions as well as air, the general ohmic relationship is used. The modelling can be carried out by FEM software such as COMSOL [6, 11].

### 3.1.7 Three-Dimensional (3D) H-Formulation Models

Similar to 2D $H$-formulation models, 3D models have three dependent variables for the magnetic field: $H_x, H_y, H_z$, as well as 3D current density: $J_x, J_y, J_z$, and electric field: $E_x, E_y, E_z$. The 3D $H$-formulation model can simulate all the conventional geometries, and it is also able to study superconductors with irregular shapes. More importantly, 3D models become indispensable for exploring the complex superconducting problem where the current and magnetic fields are not perpendicular to each other [12, 13]. Ampere's Law for the 3D model can be written as:

$$
\begin{bmatrix} J_x \\ J_y \\ J_z \end{bmatrix} = \begin{bmatrix} \dfrac{\partial H_z}{\partial y} - \dfrac{\partial H_y}{\partial z} \\[2mm] \dfrac{\partial H_x}{\partial z} - \dfrac{\partial H_z}{\partial x} \\[2mm] \dfrac{\partial H_y}{\partial x} - \dfrac{\partial H_x}{\partial y} \end{bmatrix} \tag{3.16}
$$

Substituting $H = [H_x, H_y, H_z]$ and $E = [E_x, E_y, E_z]$, Faraday's Law for the 3D model can be expressed as:

$$
\begin{bmatrix} \dfrac{\partial E_z}{\partial y} - \dfrac{\partial E_y}{\partial z} \\[2mm] \dfrac{\partial E_x}{\partial z} - \dfrac{\partial E_z}{\partial x} \\[2mm] \dfrac{\partial E_y}{\partial x} - \dfrac{\partial E_x}{\partial y} \end{bmatrix} = -\mu_0 \mu_r \begin{bmatrix} \dfrac{\partial H_x}{\partial t} \\[2mm] \dfrac{\partial H_y}{\partial t} \\[2mm] \dfrac{\partial H_z}{\partial t} \end{bmatrix} \tag{3.17}
$$

Using the norm term for $H$, $J$ and $E$:

$$
H_{norm} = \sqrt{H_x^2 + H_y^2 + H_z^2} \tag{3.18}
$$

$$
J_{norm} = \sqrt{J_x^2 + J_y^2 + J_z^2} \tag{3.19}
$$

$$
E_{norm} = \sqrt{E_x^2 + E_y^2 + E_z^2} \tag{3.20}
$$

The $E$-$J$ power law for 3D model can be defined as:

$$
\begin{bmatrix} E_x \\ E_y \\ E_z \end{bmatrix} = E_0 \begin{bmatrix} \dfrac{J_x}{J_{norm}} \left( \dfrac{J_{norm}}{J_c} \right)^n \\[3mm] \dfrac{J_y}{J_{norm}} \left( \dfrac{J_{norm}}{J_c} \right)^n \\[3mm] \dfrac{J_z}{J_{norm}} \left( \dfrac{J_{norm}}{J_c} \right)^n \end{bmatrix} \tag{3.21}
$$

According to Eqs. (3.16) to (3.21), the 3D-formulation equations could be determined with three dependent variables $H = [H_x, H_y, H_z]$, which can also be simulated by FEM software, e.g. COMSOL [14].

## 3.2 Theoretical Analysis of LFEIT

### 3.2.1 Acoustic Field Study

The most general physical parameters in ultrasound propagation analysis are: the ultrasound pressure $p$, the velocity of particle movement $v$ and the mass density of the medium, $\rho$. These three variables are dependent on Newton's 2nd Law, the Law of Mass Conservation, and the equations of State (pressure volume and temperature) [15, 16].

Starting from a small unit volume according to Newton's Law:

$$F = -s\frac{\partial p}{\partial x}dx = ma = \rho s\frac{dv}{dt}dx \tag{3.22}$$

where $s$ is the surface area of the object, $p$ is the ultrasound pressure, $\rho$ is the mass density of the medium, and $v$ is the velocity of particle movement. After simplification of Eq. (3.22), we obtain:

$$\rho\frac{dv}{dt} = -\frac{\partial p}{\partial x} \tag{3.23}$$

Adding the time-varying term of $v$ and allowing for small changes in the density $\rho_0$:

$$(\rho_0 + \Delta\rho)\left(\frac{dv}{dt} + v\frac{dv}{dx}\right) = -\frac{\partial p}{\partial x} \tag{3.24}$$

Neglecting the small second and higher order terms, Eq. (3.24) becomes:

$$\rho_0\frac{\partial v}{\partial t} = -\frac{\partial p}{\partial x} \tag{3.25}$$

According to the Law of Mass Conservation, in a certain volume the mass difference between fluids flowing in and out will equal the mass increment or decrease in this volume:

$$-\frac{\partial(\rho v)}{\partial x} = \frac{\partial p}{\partial x} = \frac{\partial\Delta\rho}{\partial t} \tag{3.26}$$

Combining Eqs. (3.25) and (3.26):

$$\rho_0\frac{\partial v}{\partial t} = -\frac{\partial\Delta\rho}{\partial t} \tag{3.27}$$

Based on the equations of State (pressure volume and temperature), with the ideal sound speed in the fluid $c_0$, the relationship between the sound pressure and mass density of medium is [13, 16]:

$$p = c_0^2 \Delta\rho \tag{3.28}$$

Rearranging Eqs. (3.25), (3.26), and (3.28), the ideal one-dimensional sound wave equation can be derived:

$$\frac{\partial^2 p}{\partial x^2} = \frac{1}{c_0} \frac{\partial^2 p}{\partial t^2} \tag{3.29}$$

Converting Eqs. (3.25) and (3.27) to three-dimensional equations:

$$\rho_0 \frac{\partial \boldsymbol{v}}{\partial t} = -\nabla p \tag{3.30}$$

$$-\nabla \cdot (\rho_0 \boldsymbol{v}) = \frac{\partial \Delta\rho}{\partial t} \tag{3.31}$$

Combining Eqs. (3.28) and (3.31), and taking the derivative of Eq. (3.30) with regard to $t$:

$$-\nabla \cdot \left( \rho_0 \frac{\partial \boldsymbol{v}}{\partial t} \right) = \frac{\partial^2 \Delta\rho}{\partial t^2} \tag{3.32}$$

Substituting Eqs. (3.28) and (3.30) into Eq. (3.32), the three-dimensional sound wave equation can be derived [13, 17]:

$$\nabla^2 p = \frac{1}{c_0} \frac{\partial^2 p}{\partial t^2} \tag{3.33}$$

where the Laplace operator is:

$$\nabla^2 = \frac{\partial^2}{\partial x^2} + \frac{\partial^2}{\partial y^2} + \frac{\partial^2}{\partial z^2} \tag{3.34}$$

Therefore, when the sound pressure has been determined, the formula for the particle movement velocity can be determined using Eqs. (3.30) and (3.33) [15]:

$$\begin{cases} v_x = -\dfrac{1}{\rho_0} \displaystyle\int \dfrac{\partial p}{\partial x} dt \\[3mm] v_y = -\dfrac{1}{\rho_0} \displaystyle\int \dfrac{\partial p}{\partial y} dt \\[3mm] v_z = -\dfrac{1}{\rho_0} \displaystyle\int \dfrac{\partial p}{\partial z} dt \end{cases} \qquad (3.35)$$

### 3.2.2 Magneto-Acousto-Electric Technique

LFEIT is based on the measurement of electrical signals arising when an ultrasound wave propagates through a conductive medium, which is subjected to a magnetic field [18, 19]. Figure 3.5 shows the schematic of a LFEIT system. According to the ionic definition of electrical conductivity, the magnitude of this Lorentz current is proportional to the amount of charges the tissue releases [18, 20]. The conductivity of the medium at the focal point can then be determined. Although these Lorentz sources occur in the interior of the domain, their currents or their corresponding electric potentials can be sampled on electrodes positioned at the boundary. This data is then used in the context of an inverse problem to reconstruct the electrical conductivity distribution throughout the domain, or locally at a certain region of interest [18, 21]. Figure 3.6 presents the flow chart for the working principle of this Magneto-Acousto-Electric technique.

According to the formula for Lorentz force:

$$\boldsymbol{F} = q\boldsymbol{v} \times \boldsymbol{B} \qquad (3.36)$$

**Fig. 3.5** Schematic of a LFEIT system [18]

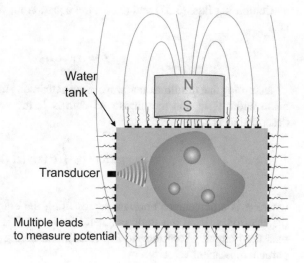

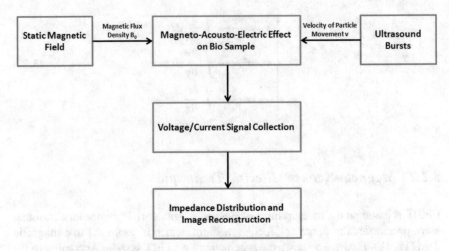

**Fig. 3.6** Flow chart for the working principle of the Magneto-Acousto-Electric technique

where $q$ is the charge of a particle moving with velocity $v$, and $\boldsymbol{B}$ is the magnetic flux density. This Lorentz force is also equivalent to the force caused by the induced electric field:

$$\boldsymbol{F} = q\boldsymbol{E} \tag{3.37}$$

Meanwhile, the current density is related to the electrical conductivity and electric field by:

$$\boldsymbol{J} = \sigma\boldsymbol{E} \tag{3.38}$$

Combining Eqs. (3.37) and (3.38), the equation for the transient current density is:

$$\boldsymbol{J} = \sigma\boldsymbol{v} \times \boldsymbol{B_0} \tag{3.39}$$

Assuming that the ultrasound wave propagates along the z direction, the ultrasound beam width is $W$ and the ultrasound path is $L$, the voltage measurement can be described as [15, 22]:

$$V_h(t) = \alpha R W B_0 \int_L \sigma(z)v(z, t)dz \tag{3.40}$$

where $\alpha$ is a percentage constant representing the efficiency current collected by the electrodes, $B_0$ is the static magnetic field, and $R$ is the total impedance of the measured circuit. According to Eq. (3.35), and taking the z direction term of the particle movement velocity:

$$v_z = -\frac{1}{\rho_0} \int \frac{\partial p}{\partial z} dt \tag{3.41}$$

The ultrasound momentum $M$ can be expressed by using the time integration of ultrasound pressure with regard to time $\tau$ [4, 15]:

$$M(z, t) = \int_{-\infty}^{t} p(z, \tau) d\tau \tag{3.42}$$

Therefore, Eq. (3.40) can be transformed by substituting Eqs. (3.41) and (3.42) [22]:

$$V_h(t) = \alpha R W B_0 \int_L M(z, t) \frac{\partial}{\partial z} \left[ \frac{\sigma(z)}{\rho(z)} \right] dz \tag{3.43}$$

It can be seen from Eq. (3.43) that the magnitude of the voltage signal is proportional to the strength of the magnetic field and the ultrasound pressure. Moreover, the voltage signal is nonzero only at the places where the gradient of electrical conductivity over mass density $\nabla(\sigma/\rho)$ is not zero. In other words, the electrical signal can be induced only at the interface between different regions within the specimen due to the Magneto-Acousto-Electric effects [15, 22].

# References

1. A. Morandi, M. Fabbri, A unified approach to the power law and the critical state modeling of superconductors in 2D, Supercond. Sci. Technol. 28(2):2015, Art. no. 024004
2. C. Sobrero, M. Malachevsky, A. Serquis, Core microstructure and strain state analysis in MgB 2 wires with different metal sheaths, Adv. Condensed Matter Phys. (2015)
3. C. Bean, Magnetization of hard superconductors. Phys. Rev. Lett. 8, 250–253 (1962)
4. Y. Kim, Critical persistent currents in hard superconductors. Phys. Rev. Lett. 9, 306–309 (1962)
5. C. Bean, Magnetization of high-field superconductors. Rev. Mod. Phys. 9, 31–38 (1964)
6. Z. Zhong, A study of critical currents and quench in 2G superconductors under varying magnetic fields, PhD Thesis, University of Cambridge (2015)
7. Y. Kim, Magnetization and critical supercurrents. Phys. Rev. Lett. 129, 528–535 (1963)
8. I. Falorio, E.A. Young, Y. Yang, Flux pinning distribution and EJ characteristics of 2G YBCO Tapes
9. J. Rhyner, Magnetic properties and AC-losses of superconductors with power law current—voltage Characterist. Physica C 212, 292–300 (1993)
10. S. Farinon, G. Iannone, P. Fabbricatore, U. Gambardella, 2D and 3D numerical modeling of experimental magnetization cycles in disks and spheres, Supercond. Sci. Technol. 27(10) (2014), Art. no. 104005
11. Z. Hong, A.M. Campbell, T.A. Coombs, Numerical solution of critical state in superconductivity by finite element software, Supercond. Sci. Technol. 19(12) (2006)
12. T. Coombs, Z. Hong, Y. Yan, C. Rawlings, The next generation of superconducting permanent magnets: The flux pumping method. IEEE Trans. Appl. Supercond. 19(3), 2169–2173 (2009)
13. M. Zhang, T. Coombs, 3D modeling of high-Tc superconductors by finite element software, Supercond. Sci. Technol. 25(1) (2011), Art. no. 015009

14. M. Zhang, Study of second generation high temperature superconducting coils, Ph.D. Thesis, University of Cambridge (2013)
15. S. Wang, 3D forward problem and experimental research of magneto-acoustic electrical tomography, Master Thesis, University of Chinese Academy of Sciences (2013)
16. J. Woo, A short history of the development of ultrasound in obstetrics and gynecology, 18 August, 2015; http://www.ob-ultrasound.net/history1.html
17. A. Montalibet, J. Jossinet, A. Matias, D. Cathignol, Electric current generated by ultrasonically induced Lorentz force in biological media. Med. Biol. Eng. Compu. **39**(1), 15–20 (2001)
18. N. Polydorides, *In-Vivo Imaging With Lorentz Force Electrical Impedance Tomography* (University of Edinburgh, UK, 2014)
19. P. Grasland-Mongrain, J. Mari, J. Chapelon, C. Lafon, Lorentz force electrical impedance tomography. IRBM **34**(4), 357–360 (2013)
20. Y. Xu, B. He, Magnetoacoustic tomography with magnetic induction (MAT-MI), Phys. Med. Biol. **50**(21) (2005)
21. H. Ammari, Y. Capdeboscq, H. Kang, A. Kozhemyak, Mathematical models and reconstruction methods in magneto-acoustic imaging. Europ. J. Appl. Math. **20**(03), 303–317 (2009)
22. H. Wen, J. Shah, R.S. Balaban, Hall effect imaging. IEEE Trans. Biomed. Eng. **45**(1), 119–124 (1998)

# Chapter 4
# Magnet Design for LFEIT

## 4.1 Introduction

Since the last century, research on the electrical impedance of human tissues has become tremendously popular all over the world [1]. The technologies to image the electrical impedance of biological tissues can make great contributions to the early diagnosis of cancer and stroke [2]. Lorentz Force Electrical Impedance Tomography (LFEIT) is a novel and promising configuration for a diagnostic scanner which is able to achieve the 3D high resolution imaging of tissue impedance based on an ultrasonically induced Lorentz force [3].

The working principle of LFEIT is to image the electrical signal generated by the magneto-acoustic effect from a biological sample [3]. As shown in Fig. 4.1, a sample is located in a magnetic field (z direction) and an ultrasound wave (y direction) propagates through its conductive medium [4]. The ultrasound propagation causes the free ions and charges in the biological tissue to move within the magnetic field. This can generate magneto-acoustic waves to induce a Lorentz current flow within the tissue [5]. LFEIT has several advantages over conventional electrical impedance tomography and other medical imaging techniques, e.g. excellent bio-detection of soft tissues, high spatial resolution, portability for emergency diagnosis (could potentially be equipped into an ambulance), and relatively low manufacturing cost [3, 4].

In a LFEIT system, the magnet is a crucial component. The intensity of the magnetic field directly affects the imaging quality of biological tissues. There are four magnet designs in Chap. 4, which are: (1) Halbach Array magnet design (perfect round shape permanent magnets), (2) Halbach Array magnet design (square shape permanent magnets), (3) Superconducting Helmholtz Pair magnet design and (4) Superconducting Halbach Array magnet design. For a small scale experimental LFEIT system, permanent magnets could be used to create an appropriate field [4].

B. Shen, *Study of Second Generation High Temperature Superconductors: Electromagnetic Characteristics and AC Loss Analysis*, Springer Theses,
https://doi.org/10.1007/978-3-030-58058-2_4

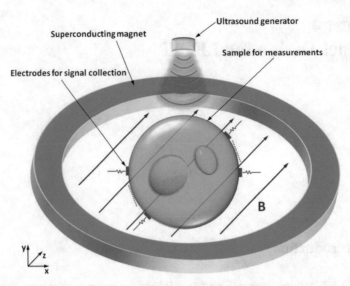

**Fig. 4.1** Configuration of superconducting Lorentz Force Electrical Impedance Tomography (LFEIT)

However, for a large diagnostic LFEIT system like a whole body scanner which requires intensity greater than 1 T, permanent magnets are not able to realize the portability for LFEIT systems due to their heavy weight and large geometry [6].

In this thesis the author tried to use a superconducting magnet in the LFEIT system. A compact superconducting magnet can produce a homogenous magnetic field with high intensity, which could be significantly beneficial for the enhancement of electrical signal output and robustness to noise, especially for a large scale LFEIT system [7].

As the most favourable structure used for MRI, a superconducting Helmholtz Pair magnet is capable of providing a magnetic field with the proper strength and extremely good homogeneity [8]. However, a Helmholtz Pair occupies a large space due to the specific arrangement for the locations of Helmholtz coils, namely that, the coils' radius $R$ is equal to the axial distance of the two coils $D$ is required to create a uniform magnetic field. A Halbach Array is an effective arrangement of magnets that is able to generate a homogenous magnetic field [9], whose geometry has the short depth in the y direction shown in Fig. 4.1 (silver colour brim). A practical Halbach magnet was tested experimentally to generate a uniform field for a Lorentz-force hydrophone [10]. Therefore, a reasonable idea is to use superconducting coils to build an electromagnet, which is essentially equivalent to the Halbach Array based on permanent magnets. This concept of a superconducting Halbach Array magnet is presented in Sect. 4.3.

Section 4.3 demonstrates the modelling of a superconducting Halbach Array magnet using the FEM software COMSOL Multiphysics. The target of this conceptual magnet design is to achieve an average magnetic strength greater than 1 T to an acceptable inhomogeneity (($B_{max}-B_{min})/B_{average}$) of less than 100 ppk (parts per thousand) within a 60 cm diameter circular cross-section for a potential full-body LFEIT system.

## 4.2 Permanent Magnet Design

Permanent magnets perform an important role in our daily lives [11]. They serve as an essential component in many power applications, instrumentation and scientific research where magnetic fields are required, such as electric motors, particle accelerators, loudspeakers and computers, etc. [6]. A high intensity magnetic field is essentially to Lorentz Force Electrical Impedance Tomography, as it can greatly improve the output signal strength and the SNR [12, 13].

There are two types of permanent magnet material which are widely used for the design of strong and uniform magnetic fields. First, Samarium-cobalt such as $SmCo_5$, a member of rare earth materials, is a good candidate for manufacturing permanent magnets [14, 15]. The second general candidate is Neodymium-iron-boron ($Nd_2Fe_{14}B$), which has been extensively used since the 1980s [16]. Both of these two types have a remnant magnetic field of typically 1–1.42 T, and coercivity of 700–1000 KA/m [11, 15, 16].

In this thesis, a Neodymium-iron-boron magnet is used for modelling the design of a permanent magnet with the remanence 1.42 T. The "Magnetic Field (mf)" physics in COMSOL AC/DC package was used, because the remnant flux density function within "Ampere's Law physics" in COMSOL is the most suitable tool for modelling permanent magnets.

### 4.2.1 Halbach Array Design Using Permanent Magnets (Perfect Round Shape)

A Halbach Array is described as a special arrangement of permanent magnets which adds magnetic field in a specific main direction while cancelling the field in other directions to near zero [10, 17]. As shown in Fig. 4.2, a typical eight-block Halbach Array is achieved using a special rotating pattern of magnetisation (blue arrows) where only the magnetic field in the y direction (red arrow) is wanted in the centre.

A 3D Halbach Array model was built using "Magnetic Field (mf)" physics model in COMSOL AC/DC module, according to the eight-block Halbach Array configuration shown in Fig. 4.2. Each single magnet occupied one eighth of the entire ring. The specification of this Halbach Array design is presented in Table 4.1.

**Fig. 4.2** Configuration of a
typical eight-block Halbach
Array

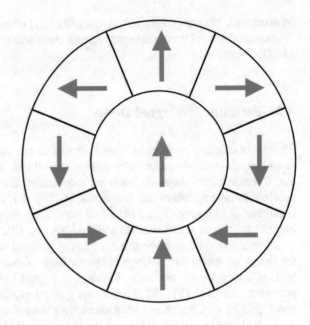

**Table 4.1** Specification for
an eight-block Halbach Array
design (idealized
configuration)

| Parameters | Value |
| --- | --- |
| Inner diameter | 60 cm |
| Outer diameter | 160 cm |
| Thickness | 15 cm |
| Remnant magnetic field for magnet | 1.42 T |

The Constitutive Relation of remnant flux density used in this design is:

$$B = \mu_0\mu_r H + B_r \qquad (4.1)$$

where $B_r$ is the remnant flux density of 1.42 T for this design.

Figure 4.3 illustrates the 3D plot of magnetic flux density of the Halbach Array
design. In the entire geometry, the maximum magnetic flux density is 2.26 T at the
joint edges of magnets at the two sides along the x-axis, while the minimum flux
density 0.03 T occurs at the top and bottom of the y-axis. Figure 4.4 presents the
zoomed in part of the centre of the Halbach ring, which is also considered to be
the most important area for this design. Figure 4.4 shows that the magnetic field is
uniform for the most part of the ring, but there are still eight high-intensity spots (up to
1.91 T) between the interfaces of two magnets, and some relatively low fields appear
besides these high-intensity field spots. Furthermore, the direction of magnetic flux
inside the Halbach ring is almost in the +y direction (the red arrow in Fig. 4.4). To be
more accurate, a 1D plot for the magnetic flux density along the x-axis diameter of the
Halbach ring is shown in Fig. 4.5. There are very sharp increments and decrements

Volume: Magnetic flux density norm (T) Arrow Surface: Direction of magnetic flux

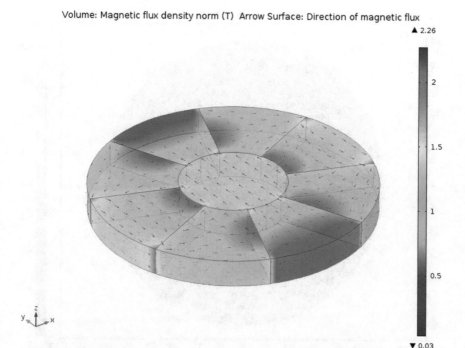

**Fig. 4.3**  3D plot for the magnetic flux density of the Halbach Array design (idealized configuration)

of the flux density at the end of two sides (from $x = -30$ cm to $x = -28$ cm, and $x = 28$ cm to $x = 30$ cm). A uniform field is achieved from $x = -20$ cm to $x = 20$ cm with an inhomogeneity of less than 50 ppk.

### 4.2.2  Halbach Array Design Using Permanent Magnets (Square Shape)

There are a few inevitable issues to be considered for the real Halbach Array design, regarding the arrangement of perfectly round permanent magnets as stated in Sect. 4.1.1. First, extremely strong forces are generated between the tiny gaps of these round shape magnets, which will possibly bring about safety issues. Second, the manufacture of round shape magnets is highly laborious and costly. Therefore, a Halbach Array normally prefers an arrangement of permanent magnets with square or rectangle shape. Figure 4.6 demonstrates a typical eight-square permanent magnet Halbach Array showing the directions of magnetisation (blue arrows). A magnetic field in the y direction (red arrow) appears in the centre of the Halbach ring.

A 3D Halbach Array model was constructed using the "Magnetic Field (mf)" physics model in the COMSOL AC/DC module, on the basis of the configuration of

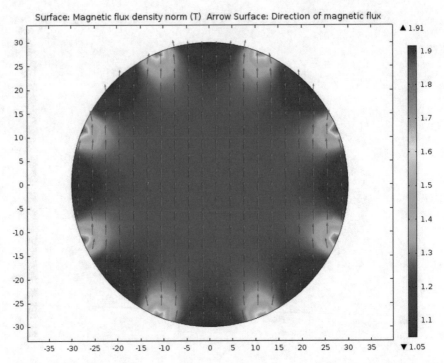

**Fig. 4.4** 2D plot for the magnetic flux density in the central circle of the Halbach Array design (idealized configuration)

the eight-square permanent magnets Halbach Array shown in Fig. 4.6. Each single magnet has a square shape, and each two have a phase angle of 90 degrees. The specification for this Halbach Array design is presented in Table 4.2.

Figure 4.7 demonstrates the 3D plot of the magnetic flux density of the Halbach Array design with square shape magnets. The maximum magnetic flux density of 1.65 T occurs at the sides of each square magnet, while the minimum flux density is close to zero at the area between each square magnet. Figure 4.8 shows the crucial part for this design—the magnified section at the Halbach ring centre. Compared to the magnetic field with perfect round shape magnets (Fig. 4.4), the magnetic field is more uniform in the Halbach array with square magnets (Fig. 4.8). There are 4 high intensity fields of 0.65 T, and some relatively lower fields appear around the sides of the circle. Similar to that shown in Fig. 4.8, the direction of magnetic flux density inside this Halbach ring is almost in the +y direction (red arrow). Figure 4.9 shows a 1D plot of the magnetic flux density along the x-axis diameter of the Halbach ring. In obvious contrast to Fig. 4.5, there is no sharp increment and decrement of the flux density at the two sides. A uniform field is achieved in the most places along the diameter with an inhomogeneity of less than 100 ppk. However, the maximum magnetic intensity is less than 0.55 T in the centre of the Halbach ring.

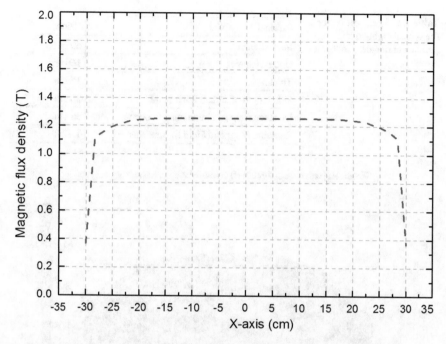

**Fig. 4.5** 1D plot for the magnetic flux density on the diameter along the x-axis of the Halbach ring

**Fig. 4.6** Configuration of a
typical eight-square
permanent magnets Halbach
Array

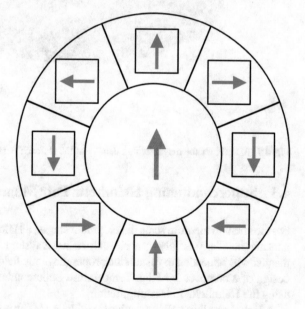

**Table 4.2** Specification for the typical eight-square permanent magnets Halbach Array

| Parameters | Value |
| --- | --- |
| Inner diameter | 60 cm |
| Outer diameter | 160 cm |
| Thickness | 15 cm |
| Side length of square magnet | 30 cm |
| Distance from square magnet to centre | 55 cm |
| Remnant Magnetic Field for magnets | 1.42 T |

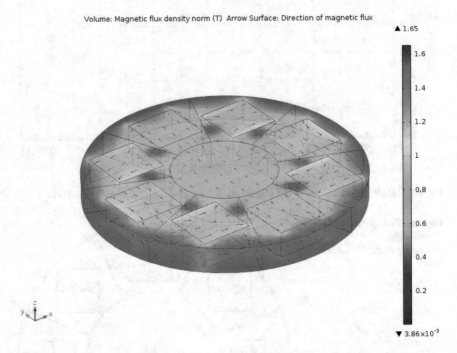

**Fig. 4.7**  3D plot for the magnetic flux density of the Halbach Array design (square shape)

## 4.3  Superconducting Helmholtz Pair Magnet Design

For the small to medium scale development stages, a Halbach Array based on permanent magnets can possibly suffice. But for a larger system, a compact superconducting magnet will be needed to sustain the proper magnetic field. This section describes the design of a compact 2G High Temperature Superconducting (HTS) electromagnet using the Helmholtz Pair configuration.

A Helmholtz Pair coil is a device to produce a highly uniform magnetic field in the centre region [8]. As shown in Fig. 4.10, A Helmholtz pair comprises two identical circular coils (radius $R$) which are symmetrically placed along a common axis. The

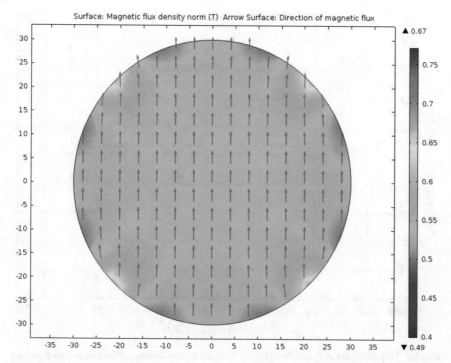

**Fig. 4.8** 2D plot for the magnetic flux density in the central circle of the Halbach Array design (square shape)

two circular coils are separated by a distance $H$, and both carry the same amount of current in the same direction. The simplest way to maximise the homogeneity of the magnetic field at the centre is to set distance $H$ equal to $R$ ($H = R$).

A 2D Helmholtz Pair model was built based on the structure of Helmholtz Pair shown in Fig. 4.10, which used the Partial Differential Equation (PDE) of the COMSOL Mathematics Module. The 2D $\boldsymbol{H}$-formulation model here uses a Cartesian coordinate model comprising 2 variables $\boldsymbol{H} = [H_x, H_y]$ as illustrated in Chap. 3. The $\boldsymbol{B}$-dependent critical current model was also used for this design. $J_c$ can be reduced in the parallel and perpendicular magnetic field [18]:

$$J_c(B) = \frac{J_0}{\left(1 + \sqrt{\frac{k^2 B_{para}^2 + B_{perp}^2}{B_0}}\right)} \tag{4.2}$$

where $J_0$ is the critical current in a zero magnetic field at 77 K. The parameters used in (4.2) are $k = 0.186$ and $B_0 = 0.426$, as presented in literature [18]. The perpendicular component is significantly higher than the parallel component.

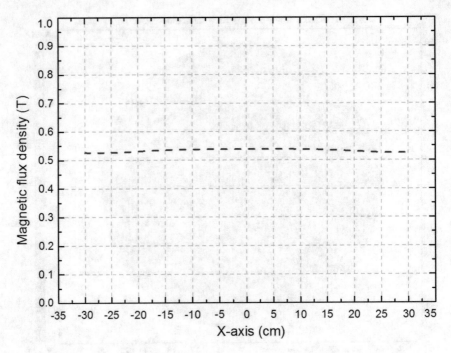

**Fig. 4.9** 1D plot for the magnetic flux density along the x-axis diameter of the Halbach ring (square shape)

**Fig. 4.10** Configuration of a Helmholtz Pair

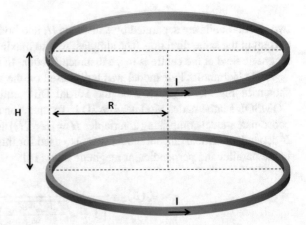

For this design, coils wound from a coated conductor were represented by continuous area bulk approximation in order to improve the model convergence and simulation speed [18]. The transport current was injected into the HTS coils using the Pointwise constraint from the general PDE model, which was able to force the integration of the current density $J_s$ within the bulk approximation over the cross-section

**Table 4.3** Specification for the superconducting Helmholtz Pair magnet design

| Parameters | Value |
|---|---|
| Inner diameter | 60 cm |
| Outer diameter | 120 cm |
| Height | 70 cm |
| $R$ | 50 cm |
| $H$ | 50 cm |
| $\mu_0$ | $4\pi \times 10^{-7}$ |
| $n$ (E-J Power Law factor) | 21 |
| $J_{c0}$ | $10^8$ A/m$^2$ |
| $E_0$ | $10^{-4}$ V/m |
| $I_{app}$ | 150 A |

area $A$ equal to the magnitude of the transport current $I_s$ in each tape multiplied by the number of turns N:

$$NI_s = \int J_s dA \tag{4.3}$$

The modelled coils was equivalent to 4800 turns ($4 \times 1200$ turns of a single layer pancake coils), of 12 mm wide YBCO coils for this design (SCS12050 SuperPower® [19], with critical 300 A at 77 K). The DC current was imposed into each coil using a ramp function with an increment of full current 150 A in the first 0.2 s, and then fixed at this value. The specification of the Helmholtz Pair design is shown in Table 4.3.

Figure 4.11 demonstrates the 3D plot of the magnetic flux density of the Helmholtz Pair magnet. Fairly high magnetic flux density appears near the HTS coils cross section, where the highest flux density reaches approximately 4.25 T. The magnetic field distribution between the upper and lower HTS coils is nearly zero, caused by the cancellation of these two fields. The rectangular box in the middle of Fig. 4.11 is the most important part for the Helmholtz Pair magnet, being where the most uniform field occurs. From the colour bar in Fig. 4.11 it can be seen that the magnitude of the uniform field is approximately 1 T, and the red arrows reveal that the direction of the uniform field is almost in the +y direction. Figure 4.12 presents the 1D plot of the magnetic flux density along the x-axis (x = −30 to x = 30 cm) of the Helmholtz Pair magnet. There are only slightly increasing and decreasing trends of the flux density at the end of two sides, and in the remaining region the magnetic field is extremely uniform (inhomogeneity less than 50 ppk).

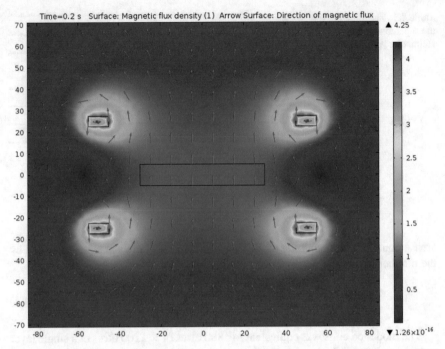

**Fig. 4.11** 2D plot for the magnetic flux density of the Superconducting Helmholtz Pair magnet design

## 4.4 Superconducting Halbach Array Magnet Design

### 4.4.1 Modelling and Simulation

As described in Sect. 4.1, a Halbach Array is a favourable arrangement of permanent magnets which can generate a proper uniform magnetic field. However, a permanent magnet based Halbach Array only can suffice for small to medium scale systems. Therefore, a reasonable idea was created that it could use superconducting coils to build an electromagnet which is essentially equivalent to the Halbach Array based on permanent magnets. This concept of a superconducting Halbach Array magnet is presented in Fig. 4.13. This section describes the design of an YBCO based HTS magnet using the Halbach Array configuration.

The superconducting Halbach Array magnet was constructed on the basis of a Halbach Array configuration with HTS coils. The modelling and simulation of this magnet was based on the partial differential equations in Sect. 3.1.6. The HTS material for this design was simulated using 12 mm wide YBCO tape manufactured by SuperPower® [19], with a critical current of 300 A at 77 K. As shown in Fig. 4.13b, the superconducting Halbach Array magnet consisted of eight HTS coils which were placed in the Halbach ring. The distances from the centre of each coil to the centre of

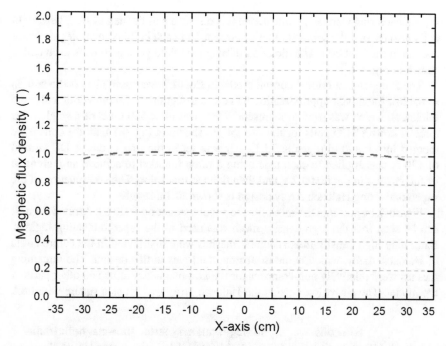

**Fig. 4.12** 1D plot for the magnetic flux density along the x-axis (x = −30 cm to x = 30 cm) of the Helmholtz Pair magnet

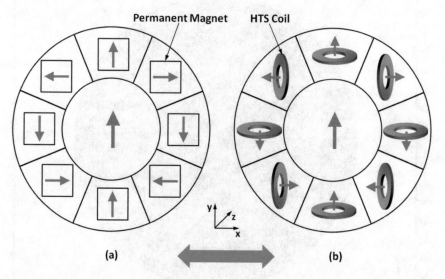

**Fig. 4.13 a** Configuration of a permanent magnet based Halbach Array magnet. **b** Configuration of an HTS coils based Halbach Array magnet

the Halbach ring were all identical, and this distance was defined as $D_c$. These eight coils also carried the same amount of current and each coil had a 90 degree phase change to the next coil. The direction of magnetic field generated from each coil is shown in Fig. 4.13b.

The $B$-dependent critical current model in Eq. (4.2) was used ($k = 0.186$ and $B_0 = 0.426$). Each bulk approximation had a cross-section of 4.8 cm width and 5 cm thickness, which was used to represent 2000 turns (4 × 500 turns of a single layer circular coil) YBCO coils for this design. As shown in Fig. 4.14, all the coils were merged into liquid nitrogen at 77 K. Air was set in the ring circle. A DC current of 120 A was applied to each tape using a ramp function to achieve an increment of full current in the first 0.2 s, and then fixed at this value. The specification of the superconducting Halbach Array design is shown in Table 4.4.

Figure 4.14 presents the mesh for the superconducting Halbach Array design. It can be seen that the high density mesh was used in the superconducting regions. Extremely high density free triangular meshes were distributed in the centre circle of Halbach Array, which is the most important part in this design. The remaining meshes were relatively non-compact for the non-magnetic frame and other parts of the model. The air region outside the Halbach array was set as a square with each

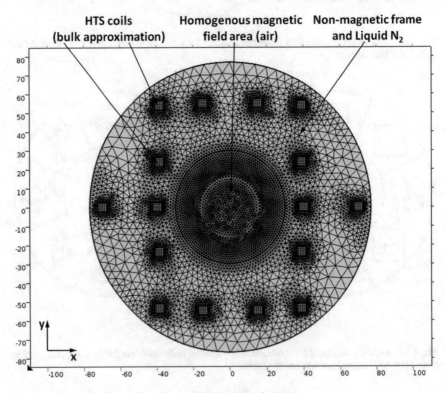

**Fig. 4.14** Mesh for the superconducting Halbach Array design

**Table 4.4** Specification for the superconducting Halbach Array design

| Parameters | Value |
|---|---|
| Inner diameter | 60 cm |
| Outer diameter | 156 cm |
| Height | 15 cm |
| HTS coil (rectangle) | Width 15 cm, Length 30 cm |
| $R_c$ (coils centre to ring centre) | 55 cm |
| YBCO tape size | Width 1.2 cm, Thickness: 0.01 cm |
| Bulk approximation cross-section | Width 4.8 cm, Length 5 cm |
| $\mu_0$ | $4\pi \times 10^{-7}$ |
| $n$ (E-J Power Law factor) | 21 |
| $J_{c0}$ | $10^8$ A/m$^2$ |
| $E_0$ | $10^{-4}$ V/m |
| $I_{app}$ | 120 A |

boundary length 10 times greater than the diameter of Halbach array, which was not shown in Fig. 4.14 for the purpose of presenting more details of this design.

Figure 4.15 illustrates the centre of superconducting Halbach ring cross-section. It can be seen that the direction of magnetic flux inside the Halbach ring centre is almost in the +y direction (the red arrow in Fig. 4.15). According to the contour plot in Fig. 4.15 for the magnetic flux density of Halbach ring centre cross-section, the flux density is around 1.0 T–1.2 T in the most part of the ring, but there are still two high-intensity areas in the end of the radius along the x-axis, while some relatively low fields appear near the top and bottom of the y-axis.

### 4.4.2 Optimization and Discussion

Some optimization can be carried out for the purpose of further improving the electromagnetic performance of the superconducting Halbach array. In this section, the optimization for the coil locations ($D_c$, the distances from each HTS coil centre to the Halbach ring centre) and different numbers of HTS coils are presented according to the simulation data.

#### A. $D_c$ Optimization

Figure 4.16 presents the relationship between the magnetic flux density on the diameter of Halbach ring centre along the x-axis and different values of $D_c$, while Fig. 4.17 shows the relationship with the y-axis. It can be found that for the cases of $D_c = 50$ cm and $D_c = 52.5$ cm, the magnetic flux density sharply rises and falls at the end of two sides of diameter along the x-axis and y-axis, which is due to the HTS

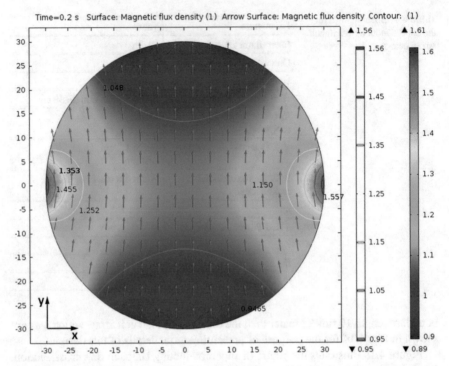

**Fig. 4.15** Magnetic flux density (with direction and contour plot) in superconducting Halbach ring centre cross-section

coils being placed too close to the Halbach ring centre. In contrast, there are only slightly decreasing and increasing trend of the flux density at the end of two sides when $D_c$ is greater than 52.5 cm. However, the magnetic flux density in most middle region along y-axis is below 1 T in the circumstances of $D_c = 57.5$ and 60 cm, which is lower than the target required.

B.   *Optimization for Numbers of Coils*

Keeping the total amount of superconducting material constant, optimization on using different numbers of coils for the Halbach arrangement can be done by shrinking each coil's size with increasing number of coils. Figure 4.18 reveals the Halbach arrangement for 12 coils (each coil has 60 degree phase change) and 16 coils (each coil has 45 degree phase change), which remains the same $D_c = 55$ cm as the 8 coil design. Figure 4.19 demonstrates the relationship between the magnetic flux density on the diameter along the x-axis for different numbers of coils. It can be seen that the magnetic field is all above 1 T for all three cases, and it becomes more uniform when the number of coils increases from 8 to 16. To be more precise, Fig. 4.20 shows the relationship between the surface inhomogeneity of the magnetic flux density in the Halbach ring centre cross-section and different numbers of coils. The magnetic homogeneity improved significantly with the increasing number of coils. With the

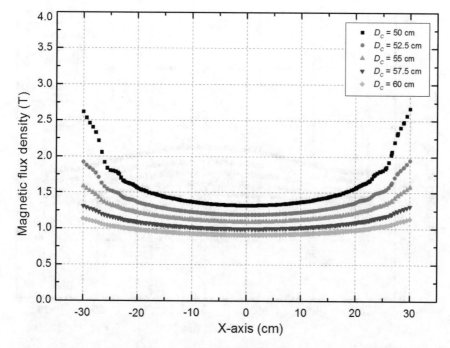

**Fig. 4.16** Relationship between the magnetic flux density on the diameter along the x-axis for different values of $D_c$ (five cases: $D_c = 50, 52.5, 55, 57.5, 60$ cm)

case of 12 coils, the inhomogeneity is better than 100 ppk (parts per thousand) inside the circular region with a 44 cm diameter for the centre cross-section. For the case of 16 coils, the inhomogeneity is below 100 ppk in the entire region of the Halbach ring centre cross-section (60 cm diameter circular region), which is broadly in the acceptable region since 100 ppk inhomogeneity is sufficient because LFEIT does not require a highly homogeneous magnetic field if the magnetic strength is around 1 T. Unlike MRI, LFEIT does not significantly rely on a highly homogenous magnetic field as LFEIT obtains the characteristic of ultrasound imaging which can realise accurate location of signals [3].

## 4.5 Summary

There are four magnet designs illustrated in Chap. 4, which are: (1) Halbach Array magnet design (perfect round shaped permanent magnets), (2) Halbach Array magnet design (square shaped permanent magnets), (3) Superconducting Helmholtz Pair magnet design and (4) Superconducting Halbach Array magnet design. Their advantages and shortcomings are discussed below.

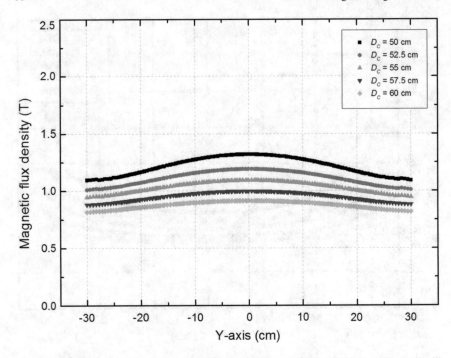

**Fig. 4.17** Relationship between the magnetic flux density on the diameter along the y-axis for different values of $D_c$ (five cases: $D_c = 50, 52.5, 55, 57.5,$ and 60 cm)

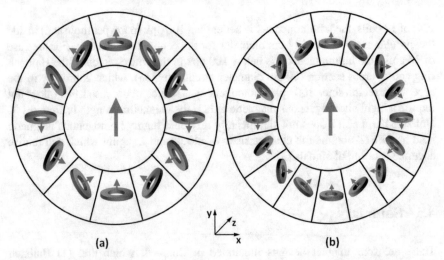

**Fig. 4.18** Halbach arrangement for: **a** 12 HTS coils, **b** 16 HTS coils, with the same values of $D_c$ as in the 8 coil design

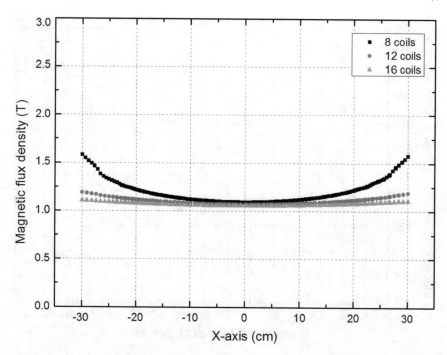

**Fig. 4.19** Relationship between the magnetic flux density on the diameter along the x-axis for different numbers of coils (three cases: 8, 12, and 16 coils)

The Halbach Array magnet design with perfect round shape permanent magnets is able to build a good magnetic field which achieves the magnetic strength greater than 1 T and with good uniformity. However, safety issues should be taken into consideration as extremely strong forces were generated between the tiny gaps of these round shape magnets. Moreover, the manufacture of the magnets is highly laborious and costly.

A Halbach Array magnet design with square shaped permanent magnets can overcome the manufacturing difficulties and reduce the mutual force generated by each magnet. The uniformity of the magnetic flux density achieved is higher than that from Halbach Array magnet design with perfect round shaped magnets. However, according to the simulation results, the biggest problem is that it cannot meet the requirement of magnetic field strength (smaller than 0.55 T).

Furthermore, both permanent magnets based Halbach Arrays are not able to realise the portability for LFEIT device due to the extremely heavy weight of Neodymium-iron-boron permanent magnets. The magnetic controllability of permanent magnet based Halbach Array is also poorer than that of electromagnets.

As the most favourable structure used for MRI, the superconducting Helmholtz Pair magnet design is capable of providing a magnetic field with the proper strength and extremely good uniformity, which has been shown in the simulation results. But the Helmholtz Pair is large due to the specific spatial arrangement for Helmholtz coils is required to create an uniform magnetic field, which could introduce difficulties

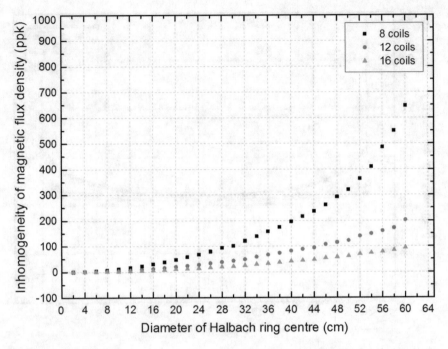

**Fig. 4.20** Relationship between the surface inhomogeneity of the magnetic flux density in the Halbach ring centre cross-section and different numbers of coils (three cases: 8, 12, and 16 coils)

on portability for a LFEIT system. Moreover, a large amount of superconducting material is needed to build the Helmholtz Pair magnet specifically for a LFEIT system. The main magnets for MRI require extremely high homogeneity, which also cannot achieve the low cost goal of the LFEIT system.

The Superconducting Halbach Array magnet design is able to overcome the disadvantages of huge size, high field homogeneity and high cost of the MRI magnet. It can generate an appropriate magnetic field within a 60 cm diameter cross-section with a thin geometry, which potentially enables the use of LFEIT for emergency diagnosis in an ambulance (such as a "Type B" ambulance). The consumption of superconducting material using in the superconducting Halbach Array magnet is approximately two-thirds of that of the superconducting Helmholtz Pair magnet, according to the data in this thesis. Analysis of the results showed that the magnetic properties such as magnitude, direction and homogeneity of the magnetic field were acceptable. Without changing the total amount of superconducting material used, the optimization results show that the magnetic homogeneity can be efficiently improved by increasing the number of coils for the Halbach arrangement while shrinking the size of each coil. The optimized case of 16 HTS coils with $D_c = 55$ cm can realize a magnetic flux density greater than 1 T with an acceptable magnetic inhomogeneity of less than 100 ppk over the entire 60 cm diameter region of the Halbach ring centre cross-section.

# References

1. C. Gabriel, S. Gabriel, E. Corthout, The dielectric properties of biological tissues: I. Literature survey, Phys. Med. Biol. **41**(11), 2231 (1996)
2. Y. Zou, Z. Guo, A review of electrical impedance techniques for breast cancer detection. Med. Eng. Phys. **25**(2), 79–90 (2003)
3. H. Wen, J. Shah, R.S. Balaban, Hall effect imaging. IEEE Trans. Biomed. Eng. **45**(1), 119–124 (1998)
4. P. Grasland-Mongrain, J. Mari, J. Chapelon, C. Lafon, Lorentz force electrical impedance tomography. IRBM **34**(4), 357–360 (2013)
5. S. Haider, A. Hrbek, Y. Xu, Magneto-acousto-electrical tomography: a potential method for imaging current density and electrical impedance. Physiol. Meas. **29**(6), 41–50 (2008)
6. J. Gieras, *Permanent Magnet Motor Technology: Design and Applications*. CRC Press (2002)
7. T. Coombs, Z. Hong, Y. Yan, C. Rawlings, The next generation of superconducting permanent magnets: the flux pumping method. IEEE Trans. Appl. Supercond. **19**(3), 2169–2173 (2009)
8. S. Lin, A. Kaufmann, Helmholtz coils for production of powerful and uniform fields and gradients, Rev. Mod. Phys. **25**(1) (1953)
9. K. Halbach, Design of permanent multipole magnets with oriented rare earth cobalt material, Nucl. Instrum. Methods **169**(1) (1980)
10. P. Grasland-Mongrain, J. Mari, B. Gilles, A. Poizat, J. Chapelon, C. Lafon, Lorentz-force hydrophone characterization, IEEE Trans. Ultrason. Ferroelect. Freq. Control **61**(2), 353–363 (2014)
11. E. Furlani, *Permanent Magnet and Electromechanical Devices: Materials, Analysis, and Applications*. Academic Press (2001)
12. N. Polydorides, *In-Vivo Imaging With Lorentz Force Electrical Impedance Tomography* (University of Edinburgh, UK, 2014)
13. P. Grasland-Mongrain, R. Souchon, F. Cartellier, A. Zorgani, J.-Y. Chapelon, C. Lafon, S. Catheline, Imaging of shear waves induced by lorentz force in soft tissues, Phys. Rev. Lett. **113**(3) (2014), Art. no. 038101
14. T. Annis, J. Eberly, Electricity generating apparatus utilizing a single magnetic flux path, Google Patents (2007)
15. Y. Ito, K. Yasuda, R. Ishigami, S. Hatori, O. Okada, K. Ohashi, S. Tanaka, Magnetic flux loss in rare-earth magnets irradiated with 200 MeV protons. Nucl. Instrum. Methods Phys. Res., Sect. B **183**(3), 323–328 (2001)
16. X. Du, T. Graedel, Global rare earth in-use stocks in NdFeB permanent magnets. J. Ind. Ecol. **15**(6), 836–843 (2011)
17. S. Jang, S. Jeong, D. Ryu, S. Choi, Design and analysis of high speed slotless PM machine with Halbach array. IEEE Trans. Magn. **37**(4), 2827–2830 (2001)
18. M.D. Ainslie, Y. Jiang, W. Xian, Z. Hong, W. Yuan, R. Pei, T.J. Flack, T.A. Coombs, Numerical analysis and finite element modelling of an HTS synchronous motor. Phys. C, Supercond. **470**(20), 1752–1755 (2010)
19. SuperPower, "SuperPower® 2G HTS Wire Specifications," Schenectady, NY 12304, USA (2014)

# Chapter 5
# Optimization of the Superconducting Halbach Array

## 5.1 Introduction

In Chap. 4, a compact superconducting magnet for use with Lorentz Force Electrical Impedance Tomography (LFEIT) has been proposed. High Temperature Superconductor (HTS) was used to design an electromagnet that performs equivalently to a permanent magnet based Halbach Array. The simulations of the superconducting Halbach Array were carried out using $H$–formulation based on $B$–dependent critical current density and bulk approximation, with the FEM platform COMSOL Multiphysics.

The superconducting Halbach Array is able to achieve an average magnetic flux density greater than 1 T and acceptable inhomogeneity within a 60 cm diameter circular cross-section for a potential full-body LFEIT system. The quality of the magnetic field, e.g. the strength and homogeneity, directly affect the imaging quality of biological tissues for a LFEIT system [1, 2]. In previous results, it has been deduced that the magnetic homogeneity can be improved by increasing the number of coils for the Halbach arrangement while shrinking each coil's size, without changing the overall total amount of superconducting material used [3].

This Chapter presents the further optimization of the coil locations and the number of coils, to obtain a better homogeneity of the magnetic field in the centre area of the superconducting Halbach Array, on the premise of maintaining the total amount of superconducting material used. Furthermore, the mathematical expressions are derived for inhomogeneity estimation versus numbers (up to infinity) of HTS coils and distance to the Halbach ring centre.

B. Shen, *Study of Second Generation High Temperature Superconductors: Electromagnetic Characteristics and AC Loss Analysis*, Springer Theses, https://doi.org/10.1007/978-3-030-58058-2_5

## 5.2  Modelling for Optimization

Modelling of the superconducting Halbach was on the basis of the partial differential equations of 2D $H$–formulation described in Chap. 3. The 12 mm HTS tape with a critical current of 300 A at 77 K, SuperPower SCS12050, was used as the coil material. As shown in Fig. 5.1a, b, and c, the superconducting Halbach Array magnet consisted of 8, 12 and 16 HTS coils, which followed the rule of 4 N (N is the integer starting from 2) for the numbers of coils. Each coil carried the same amount of current. Each coil had a 360/2 N° phase change to the next coil, for example: 90° for the 8-coil structure, 60° for the 12-coil structure, and 45° for the 16-coil structure. The ideal direction of the magnetic field generated from each coil is shown in Fig. 5.1.

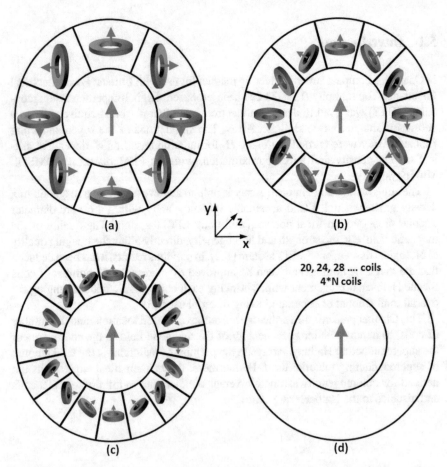

**Fig. 5.1** Configuration of the superconducting Halbach Array: **a** 8-coil, **b** 12–coil, **c** 16-coil, **d** increasing numbers of coils 4 N (N is the integer starting from 2)

**Table 5.1** Specification for optimizing the superconducting Halbach Array magnet

| Parameters | Value |
|---|---|
| Inner diameter | 60 cm |
| Outer diameter | 156 cm |
| $D_c$ (distance from coils center to ring center) | 55 cm |
| YBCO tape cross-section | 12 mm $\times$ 0.1 mm |
| Bulk approximation cross-section | 48 mm $\times$ ratio of turns |
| $\mu_0$ | $4\pi \times 10^{-7}$ H/m |
| $n$ (E–J power law factor) | 21 |
| $J_{c0}$ | $2.5 \times 10^8$ A/m$^2$ |
| $E_0$ | $10^{-4}$ V/m |
| $I_{app}$ | 120 A |
| N (number of coils) | $4 \times$ n (n is the integer starting from 2) |

The specification of the superconducting Halbach Array design is shown in Table 5.1. An example of the mesh for a 20-coil superconducting Halbach Array is shown in Fig. 5.2 (the hidden air region outside the Halbach array was more than 100 times bigger).

As the critical current of the HTS tape has anisotropic characteristics, the **B**-dependent critical current model was:

$$J_c(B) = \frac{J_{c0}}{\left(1 + \sqrt{\frac{k^2 B_{para}^2 + B_{perp}^2}{B_0^2}}\right)} \tag{5.1}$$

where $J_{c0}$ is the critical current density in a self-field, at 77 K. In Eq. (5.1), $k = 0.186$ and $B_0 = 0.426$.

Similar to the previous study, stacked coils with different layers of substrate and normal materials were represented by a continuous area bulk approximation [4, 5], which can efficiently improve the simulation and model convergence [6, 7]. The Physics function Pointwise constraint from the general PDE model was used to inject the transport current into the HTS coils, which forced the multiplication of the transport current $I_s$ by the number of turns N equal to the integration of the current density $J_s$ within the bulk approximation over the cross-section area $A$:

$$N I_s = \int J_s dA \tag{5.2}$$

Figure 5.2 shows that the coils were cooled by liquid nitrogen at 77 K. An overall DC current of 120 A for a single tape was applied using a ramp function to reach

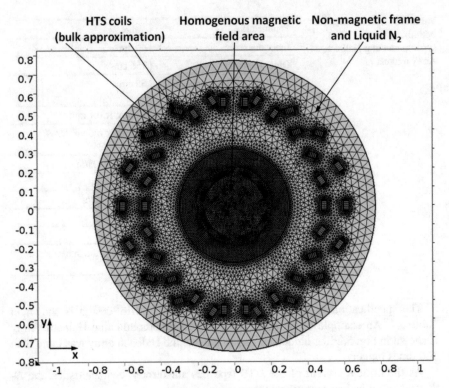

**Fig. 5.2** Mesh for a 20-coil superconducting Halbach Array design

the peak current in the first 0.2 s, and then maintained at this DC value. Each bulk approximation had a cross-section of 48 mm width, which represents 4 vertical stack coils (4 × 12 mm). In the bulk approximation, thickness corresponded to the number of turns so that the greater the thickness, the more turns.

## 5.3 Optimization and Discussion

The optimization was performed on the basis of a constant length of superconductor. Increasing the number of coils for the Halbach arrangement was achieved by shrinking the size of each coil.

Figure 5.3 presents the most crucial part of this design, the 2D centre cross-section of the superconducting Halbach with six cases: (a) 8-coil, (b) 12-coil, (c) 16-coil, (d) 20–coil, (e) 24-coil, and (f) 28-coil structures. It can be seen that for all six cases the direction of magnetic flux is almost in the +y direction (the red arrow in Fig. 5.3), but the field direction is closer to the +y direction when the number of coils is increased from 8 to 28, which is more obvious at the two ends of the x-axis. According to the surface and contour plot in Fig. 5.3, the flux density is relatively

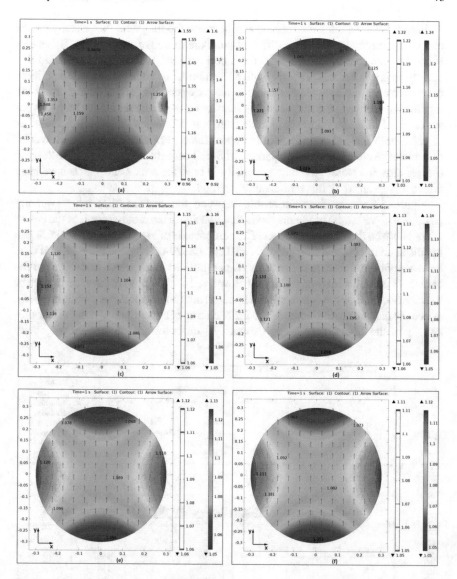

**Fig. 5.3** Magnetic flux density (with direction and contour plot) in superconducting Halbach ring center cross-section: **a** 8-coil, **b** 12-coil, **c** 16-coil, **d** 20–coil, **e** 24-coil, **f** 28-coil

uniform for the most part of the ring, but there are still two high-intensity areas in the end of the radius along the x-axis. From the colour bars of Fig. 5.3a–f, it can be seen that the difference between the maximum and minimum of the magnetic flux density ($B_{max} - B_{min}$) decreases steadily from 0.68 to 0.07 T, which implies that the magnetic field in the cross-section becomes more homogeneous.

Figure 5.4 illustrates the relationship between the magnetic flux density on the diameter along the x-axis with different numbers of coils from 8 to 28 (six cases). It can be seen that in the end sides the 8-coil case has much higher magnetic flux density, and this decreases sharply when the number of coils increases. The middle parts of the curves have similar value in all six cases. Figure 5.5 demonstrates the relationship between the magnetic flux density on the diameter along the y-axis for six cases of different numbers of coils from 8 to 28. When increasing the number of coils, both the end and the middle part of the curves become flat, which can be clearly discovered from the comparison of 8-coil case (black curve) and the 28-case (purple curve). Overall, the inhomogeneity in both the x-axis and y-axis have been effectively reduced by this optimization method.

Figure 5.6 shows the relationship between the surface (2D) inhomogeneity $((B_{max}-B_{min})/B_{average})$ of the magnetic flux density in the cross-section of the centre of the Halbach ring for the six cases. The vertical axis of Fig. 5.6 is a log-scale. When the distance to Halbach ring centre is <10 cm, the 24-coil case and the 28-coil case have an inhomogeneity to the ppm (parts per million) level, which is $<10^{-3}$. For the case of 28 coils, the inhomogeneity is below $5 \times 10^{-2}$ (50 ppk) in the entire region of the cross-section of the centre of the Halbach ring (60 cm diameter circular region). When the distance to the centre of the Halbach ring is larger than 10 cm, the inhomogeneity of the 28-coil case is approximately 1 order lower than the inhomogeneity

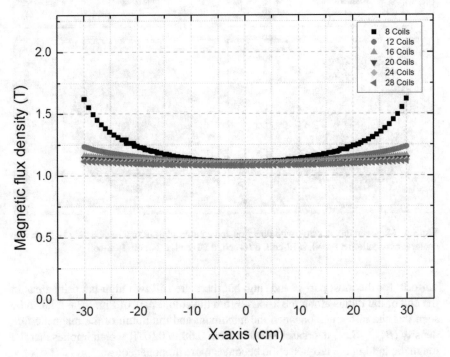

**Fig. 5.4** Relationship between the magnetic flux density on the diameter along the x-axis for different numbers of coils (six cases: 8, 12, 16, 20, 24, and 28 coils)

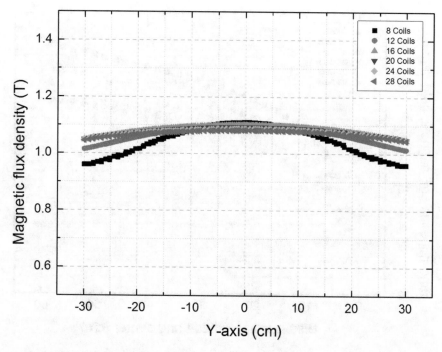

**Fig. 5.5** Relationship between the magnetic flux density on the diameter along the y-axis for different numbers of coils (six cases: 8, 12, 16, 20, 24, and 28 coils)

of the 8-coil case. The magnetic homogeneity has improved significantly with the increasing number of coils, but still using the same total amount of superconducting material.

In order to further optimize the magnetic homogeneity in the cross-section of the centre of the Halbach ring, the mathematical relationship between the inhomogeneity and increasing numbers of coils has been investigated. The equation fitting results for the relation between the inhomogeneity, distance to the centre of the Halbach ring and the number of coils are:

$$Inhomogeneity = \alpha \cdot D^{\beta} \qquad (5.3)$$

$$\alpha = a \cdot \exp(b \cdot n) + c \qquad (5.4)$$

$$\beta = e \cdot \exp(f \cdot n) + g \qquad (5.5)$$

Equation (5.3) reveals that the inhomogeneity has a power law relationship with $D$ (distance to the centre of the Halbach ring). $\alpha$ and $\beta$ are the variables for the power law which relate to n (number of coils) in Eqs. (5.4) and (5.5). Both $\alpha$ and $\beta$

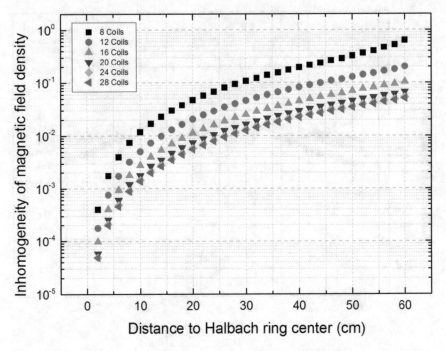

**Fig. 5.6** Relationship between the surface (2D) inhomogeneity of the magnetic flux density in the cross-section of the centre of the Halbach ring for different numbers of coils (six cases: 8, 12, 16, 20, 24, and 28 coils)

have an exponential relationship with n, and adding a constant, where $a = 1.408E$-4, $b = -0.1556$, $c = 1.188E$-05, $e = 43.32$, $f = -0.5279$, and $g = 2.022$.

From Eqs. (5.3) to (5.5), it can be deduced that although homogeneity can be improved by increasing the number of coils, the efficiency of homogeneity enhancement will be weakened when n (numbers of coils) increases to a high value. For example, for $D = 60$ cm, the inhomogeneity of the 32-coil case is approximately one order smaller than the 8-coil case, but from the mathematical estimation the inhomogeneity of the 64-coil case is only about 10% smaller than the 32-coil case, and it is estimated that further increasing the number of coils becomes much less effective at reducing the inhomogeneity.

Therefore, it can be concluded that a superconducting Halbach Array could potentially achieve a homogeneous magnetic field to the ppm level over a diameter <1/15 of its entire geometry (<10 cm distance to the centre of the Halbach ring), but it is difficult for a superconducting Halbach Array to realise the magnetic inhomogeneity to the ppm level over the entire region of the cross-section of the centre of the Halbach ring (60 cm diameter circular region) with this method. Another potential concern is that the returning field lines associated with the actual thin depth could further slightly affect the inhomogeneity in the actual 3D geometry. However, for a LFEIT system a compact superconducting Halbach Array is still usable, because 100 ppk inhomogeneity is sufficient. This is because LFEIT obtains the characteristic of ultrasound imaging to realise the accurate location of signals, and does not require a highly homogeneous magnetic field if the magnetic strength is around 1 T [2, 3]. A Halbach Array configuration based superconducting magnet could still be a feasible solution for generating a uniform magnetic field with relatively compact geometry.

## 5.4 Summary

The concept of a Halbach Array structure based superconducting magnet has been proposed to generate an appropriate homogeneous magnetic field. The simulation was executed on the FEM platform of COMSOL Multiphysics, which consists of 2D models of a superconducting Halbach Array magnet using $H$-formulation based on $B$-dependent critical current density and bulk approximation. Optimization of the magnetic homogeneity was carried out to increase the number of coils for the Halbach arrangement while shrinking the size of each coil, but still maintaining the total amount of superconducting material. The mathematical relationship between the inhomogeneity and increasing numbers of coils derived using optimization results, which demonstrated that the magnetic homogeneity can be improved by increasing the number of coils, but the efficiency for homogeneity improvement will decrease when the number of coils increases to a high value. Overall, with this optimization method, a Halbach Array configuration based superconducting magnet can potentially generate a uniform magnetic field greater than 1 T with inhomogeneity to the ppm level, which also has the advantage of relatively compact geometry. However, the fabrication difficulty and the optimization efficiency must be taken into consideration.

## References

1. P. Grasland-Mongrain, J. Mari, J. Chapelon, C. Lafon, Lorentz force electrical impedance tomography. IRBM **34**(4), 357–360 (2013)
2. B. Shen, L. Fu, J. Geng, X. Zhang, H. Zhang, Q. Dong, C. Li, J. Li, T. Coombs, Design and simulation of superconducting Lorentz force electrical impedance tomography (LFEIT). Phys. C, Supercond. **524**, 5–12 (2016)
3. B. Shen, L. Fu, J. Geng, H. Zhang, X. Zhang, Z. Zhong, Z. Huang, T. Coombs, Design of a superconducting magnet for Lorentz force electrical impedance tomography. IEEE Trans. Appl. Supercond. **26**(3), Art. no. 4400205 (2016)
4. V.M. Zermeno, A.B. Abrahamsen, N. Mijatovic, B.B. Jensen, M.P. Sørensen, Calculation of alternating current losses in stacks and coils made of second generation high temperature superconducting tapes for large scale applications. J. Appl. Phys. **114**(17), Art. no. 173901 (2013)
5. J. Xia, H. Bai, J. Lu, A. V. Gavrilin, Y. Zhou, H. W. Weijers, Electromagnetic modeling of REBCO high field coils by the H-formulation. Supercond. Sci. Technol. **28**(12) (2015)

6. L. Quéval, V. M. Zermeño, F. Grilli, Numerical models for ac loss calculation in large-scale applications of HTS coated conductors. Supercond. Sci. Technol. **29**(2) (2016)
7. M.D. Ainslie, Y. Jiang, W. Xian, Z. Hong, W. Yuan, R. Pei, T.J. Flack, T.A. Coombs, Numerical analysis and finite element modelling of an HTS synchronous motor. Phys. C, Supercond. **470**(20), 1752–1755 (2010)

# Chapter 6
# LFEIT System Design

## 6.1 Introduction

Lorentz Force Electrical Impedance Tomography (LFEIT), also known as Hall Effect Imaging (HEI) or Magneto-Acousto-Electric Tomography (MAET), is one of the most promising hybrid portable devices with burgeoning potential for cancer and internal haemorrhage detection [1–3]. In contrast to the disadvantage of ultrasonic imaging technology in distinguishing between soft tissues because acoustic impedance varies by less than 10% among muscle and blood, LFEIT shows its powerful capability for providing information about the pathological and physiological condition of tissues because electrical impedance varies widely among soft tissue types and pathological states [4]. Other than carcinomas, tissues under haemorrhage or ischemia conditions can exhibit huge difference in electrical properties because most body fluids and blood have fairly different permittivities and conductivities compared to other soft tissues [5]. Figure 6.1 presents a schematic of a superconducting Lorentz Force Electrical Impedance Tomography (LFEIT) system. LFEIT is based on electrical signals which arise when an ultrasound wave propagates through a conductive medium subjected to a magnetic field [1]. The magnitude of the electrical signal is proportional to the strength of the magnetic field and the pressure of the ultrasound wave [1].

The historical development and review of LFEIT was presented in Chap. 2. As the magnetic field is crucial for generating the final electrical signal in LFEIT, the author have tried a novel combination of superconducting magnets with a LFEIT system. The reason for this is that superconducting magnets can generate a magnetic field with high intensity and homogeneity [6], which could efficiently enhance the electrical signal induced from the sample and thus improve the signal-to-noise ratio (SNR), particularly for a large scale full-body LFEIT. This chapter presents the

B. Shen, *Study of Second Generation High Temperature Superconductors: Electromagnetic Characteristics and AC Loss Analysis*, Springer Theses, https://doi.org/10.1007/978-3-030-58058-2_6

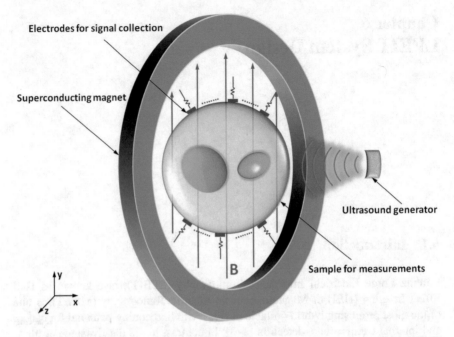

**Fig. 6.1** Schematic of superconducting Lorentz Force Electrical Impedance Tomography (LFEIT)

simulation of a superconducting LFEIT system, which includes the modelling of a superconducting magnet using the FEM software COMSOL Multiphysics, coupled with the mathematical model of the magneto-acoustic effect from LFEIT.

## 6.2  Acoustic Module Design

The acoustic module is one of the crucial parts for LFEIT. The acoustic module generates ultrasound waves causing localised mechanical vibrations in a conductive tissue located in a static magnetic field, from which Hall voltages are induced.

Figure 6.2 demonstrates the simulation of an ultrasound pressure field based on the ultrasound Matlab package from FOCUS [7]. This acoustic module used an ultrasound phase array structure, consisting of 32 transducer elements, each generating a 1 MHz ultrasound signal. The focused region had 1 mm width and 10 cm length, and the focused ultrasound pressure was approximately 3 MPa. The location of the ultrasound focused point could be tuned by changing the phase delay of each transducer. The speed of the ultrasound was set as the average value 1500 m/s in human tissues. The attenuation coefficient was set as $8 \times 10^{-2}$ dB/cm/MHz, which is close to the value for human muscle.

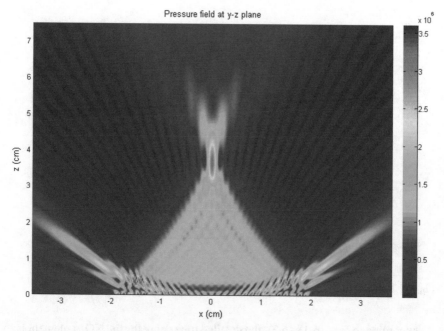

**Fig. 6.2** Simulation of an ultrasound pressure field in the y–z plane

## 6.2.1 Focusing Method of Ultrasound Phase Array

An ultrasonic phased array has the advantage that it is able to realise software-control focusing and beam deflection, with the main approach of controlling the time delay of various signals [8].

There are $2n$ transducers in the array in Fig. 6.3. The distance between adjacent transducers is $d$, the distance between the axis centre to the focus point is $F$, and the deflection angle of the beam centre is $\theta$. The triangle of OPB is shown in Fig. 6.3, according to the trigonometric theorem:

$$PB^2 = F^2 + (nd)^2 + 2nFd \sin \theta \tag{6.1}$$

The acoustic path difference between NO. $n$ element to the centre array:

$$\Delta S = F - PB \tag{6.2}$$

Assuming that the sound velocity is $c$ and a constant which avoids the negative delay $t_0$, it can be deduced that the delay value for the NO. $n$ element is [8]:

$$t_n = \frac{\Delta S}{c} + t_0 \tag{6.3}$$

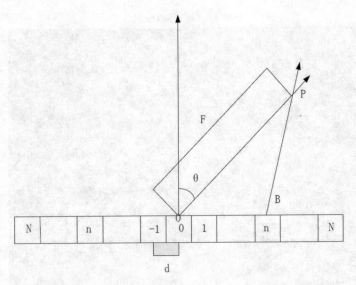

**Fig. 6.3** Focusing of ultrasonic phased array [8]

According to Eqs. (6.1)–(6.3), the detailed time delay for the NO. $n$ element is:

$$t_n = \frac{F}{C}\left[1 - \sqrt{1 + \left(\frac{nd}{F}\right)^2 - 2\frac{nd\sin\theta}{F}}\right] + t_0 \tag{6.4}$$

### 6.2.2   *Ultrasound for LFEIT*

One of the biggest advantages of LFEIT is it does not significantly rely on a highly homogenous magnetic field. This situation is quite different to Magnetic Resonance Imaging. MRI is very dependent on magnetic homogeneity, because the receiving RF signals cover the information of nuclei location whose signal magnitude is determined by the main magnetic field in the MRI. In other words, in MRI the nonuniformity of the magnetic field could cause misallocation of the original signal for imaging. Thus the magnetic uniformity of MRI directly relates to the imaging quality, and poor magnet homogeneity can cause various problems including spatial distortion, blurring, shading, etc. A typical 3.0 T magnet for MRI usually guarantees an inhomogeneity of <1 ppm over a 40 cm DSV. However, LFEIT can use the focusing area of ultrasound to identify the location of the signal, and the focusing of ultrasound is able to confer Electrical Impedance Tomography with the characteristic of ultrasound imaging, whose spatial precision is close to the resolution of ultrasound imaging.

The simulation results in the following sections will show that the quality of signal imaging was still acceptable when more than 10% inhomogeneity in the static magnetic field was applied to LFEIT system, as long as the magnitude of the magnetic flux density is larger than 1 T. The tolerance of a LFEIT system to magnetic field inhomogeneity is several orders high than that of MRI.

## 6.3   Electrical Signal Simulation

It has been described in Chap. 3 that the governing equation for the final output signal of LFEIT can be defined as:

$$V_h(t) = \alpha R W B_0 \int_L M(z, t) \frac{\partial}{\partial z} \left[ \frac{\sigma(z)}{\rho(z)} \right] dz \qquad (6.5)$$

Assuming that the ultrasound wave propagates along the z direction, $W$ is the ultrasound bean width, $L$ is the ultrasound path, $\alpha$ is a percentage constant representing the efficiency current collected by the electrodes, $B_0$ is the static magnetic field, and $R$ is the total impedance of the measurement circuit. The ultrasound momentum $M$ can be expressed using the time integration of ultrasound pressure $p$ with regard to time $\tau$:

$$M(z, t) = \int_{-\infty}^{t} p(z, \tau) d\tau \qquad (6.6)$$

Equations (6.5) and (6.6) reveal that the magnitude of the final output signal is proportional to the strength of the magnetic field and the ultrasound pressure. More importantly, the output signal is nonzero only at the interface where the gradient of electrical conductivity over mass density $\nabla(\sigma/\rho)$ is not zero. The mathematical Matlab model of a LFEIT system was built based on governing Eqs. (6.5) and (6.6).

For this model, it was assumed on the basis of the literature [3, 9, 10] that the gradient of electrical conductivity over mass density at the interface between two materials is in a second-order form:

$$\frac{\partial}{\partial z}(\sigma/\rho) = a\left(c - (z - b)^2\right) \qquad (6.7)$$

Therefore, the ideal output (absolute value) detected from the sample should be also in a second-order form with an absolute uniform magnetic field of 1 T and zero noise, as shown in Fig. 6.4.

The sample shown in Fig. 6.4 was simulated as a human tissue immersed into oil. This sample had a round shape cross-section whose diameter was 8 cm, and is

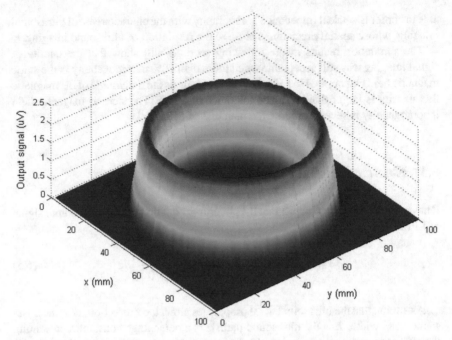

**Fig. 6.4** Ideal output signal (absolute value) detected from a sample with absolute uniform magnetic field 1 T and zero noise

located in an absolutely uniform magnetic field 1 T with zero noise. The gradient of electrical conductivity over mass density at the interface $\nabla(\sigma/\rho)$ was set as: $a = 0.03$, $b = 6$ and $c = 36$ for Eq. (6.7). It can be seen from Fig. 6.4 that the absolute value of the output signal matches the overshot of a second-order form. The output signal level is to the 1 $\mu$V order, and the peak quantity is around 2.5 $\mu$V according to the simulation.

The output signal will be distorted if the magnetic field in the testing area has 30% inhomogeneity, as shown in Fig. 6.5. Based on the literature [3], the noise signal level for LFEIT or HEI experiment is around 1 $\mu$V for a 1 MHz input ultrasound signal. Thus, a white Gaussian noise with rms value 1.5 $\mu$V was added into the entire testing region of this model, as shown in Fig. 6.6.

## 6.4   Imaging of Electrical Signal and Comparison

After obtaining the electrical signal distribution for the cross-section of the biological sample, the fundamental method for imaging can be used to determine the location and external shape of the biological sample, as an electrical signal can only be induced at the different interfaces within the sample due to the physical principle of LFEIT. The basic imaging method with the "imshow" command of MATLAB

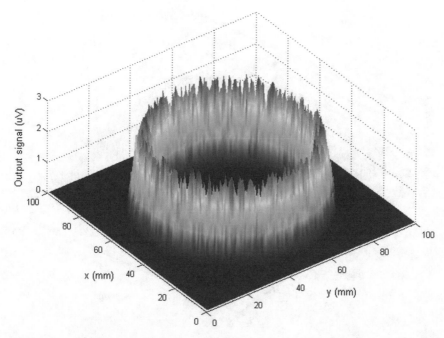

**Fig. 6.5** Output signal (absolute value) detected from a sample located in a magnetic field with 30% inhomogeneity

was used to map the electrical signals. There were two biological samples used for simulation, and both were assumed to be fully immersed into oil. The first sample had the gradient of electrical conductivity over mass density $\nabla(\sigma/\rho)$ of Eq. (6.7) with coefficients $a = 0.03$, $b = 5$ and $c = 25$, and the second sample had the $\nabla(\sigma/\rho)$ of Eq. (6.7) with coefficients $a = 0.1$, $b = 2$ and $c = 4$. Similarly, the entire testing region was simulated with a white Gaussian noise whose rms value was 1.5 μV, and the source of ultrasound was assumed to be unchanged. The magnetic field properties from different magnet designs (in Chaps. 4 and 5) were imported into the LFEIT model. Figures 6.7, 6.8, and 6.9 present the simulation for these two samples tested by the LFEIT system, with different magnetic field strengths and uniformities.

Figure 6.7 illustrates the voltage output and the electrical signal imaging from the samples located in the 0.5 T magnetic field with 30% inhomogeneity (using the 8-coil magnet with 60 A current, and using the field area near the edge of the circular region d = 50 cm). It can be discovered from Fig. 6.7a that both of the voltage outputs (absolute value) were distorted due to the inhomogeneity of the magnetic field. Figure 6.7b presents output signals which were almost submerged by the noise because the magnitude of the voltage outputs declined after distortion, and the output

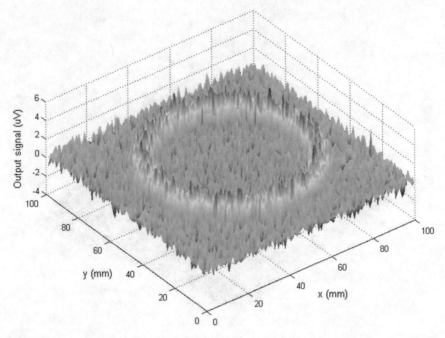

**Fig. 6.6** Output signal detected from a sample located in a 30% inhomogeneity magnetic field with noise condition

signals from the second sample were almost below 1.5 μV. The average SNR was only 3.6 dB. The image reconstruction of the electrical signal is faint and it is very difficult to find the edge and location of the second sample.

In contrast with Figs. 6.7 and 6.8 presents a magnetic field with higher uniformity (10% inhomogeneity) which was applied to the LFEIT system, but the strength of magnetic field still remained the same (using the 12-coil magnet with 60 A current, and using the field area near the edge of the circular region d = 50 cm). Compared to Fig. 6.7a, it can be found that both of the voltage outputs (absolute value) had less distortion due to the better uniformity of the magnetic field. As shown in Fig. 6.8b, although the average SNR improved to 5.6 dB, most output signals from the second sample were still missing after adding the noise. The quality of electrical signal imaging from Fig. 6.8 is slightly better than that of Fig. 6.7, and the boundary of the second sample can be hazily seen in Fig. 6.8c.

Figure 6.9 demonstrates the simulation of output signals and the electrical signal imaging where the sample was tested in the 1 T magnetic field with 30% inhomogeneity (using the 8-coil magnet with 120 A current, and using the field area near the edge of the circular region d = 50 cm). The strength of the magnetic field was doubled compared to the previous two simulations. It can be seen from Fig. 6.9b that most voltage outputs from both the samples were greater than the average noise

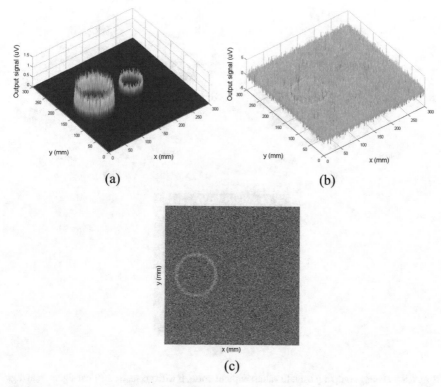

**Fig. 6.7  a** voltage output (absolute value) without noise, **b** with noise, **c** electrical signal imaging: from the sample located in the 0.5 T magnetic field with 30% inhomogeneity

level (average SNR 11.8 dB), although voltage outputs were distorted by 30% inhomogeneity of the applied magnetic field. As a result, the imaging of the electrical signals from Fig. 6.9c is much clearer than that from Figs. 6.7c to 6.8c, and the edge and shape of both samples can be discovered.

## 6.5  Summary

A mathematical model for the LFEIT system was constructed, based on the magneto-acoustic effect together with the magnetic properties from the modelling of the super-conducting Halbach Array from previous chapters. Two samples located in three different magnetic fields were simulated by the LFEIT model. According to the simulation results, both the uniformity and the strength of the magnetic field affected the electrical signal from the LFEIT system. By contrast, increasing the magnetic

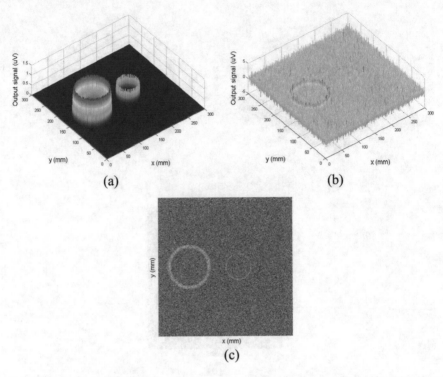

**Fig. 6.8** **a** voltage output (absolute value) without noise, **b** with noise, **c** electrical signal imaging: from the sample located in the 0.5 T magnetic field with 10% inhomogeneity

intensity is more efficient, particularly if the bio-induced electrical signal is lower than or comparable to the noise level. The performance of signal imaging is still acceptable when more than 10% inhomogeneity of magnetic flux density (around 1 T) is applied to the LFEIT system. The combination of a superconducting magnet with the LFEIT system is a reasonable approach because superconducting magnets are able to produce magnetic field with high intensity and good homogeneity, which could potentially enhance the SNR of the LFEIT system and the quality of biological imaging.

The tolerance to magnetic field inhomogeneity for the LFEIT system is several orders higher than that of MRI, due to LFEIT sharing the characteristic of ultrasound imaging [1]. The design above simulated and imaged the electrical signals generated by LFEIT, which could be used to find the approximate location and the boundary shape of a certain tissue. Nevertheless, for detailed imaging of impedance distribution, future work is required to be carried out.

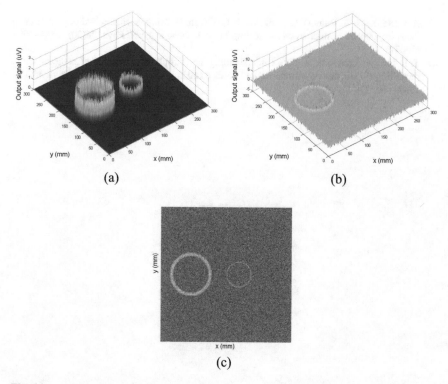

**Fig. 6.9  a** voltage output (absolute value) without noise, **b** with noise, **c** electrical signal imaging: from the sample located in the 1 T magnetic field with 30% inhomogeneity

## References

1. P. Grasland-Mongrain, J. Mari, J. Chapelon, C. Lafon, Lorentz force electrical impedance tomography. IRBM **34**(4), 357–360 (2013)
2. Y. Xu, B. He, Magnetoacoustic tomography with magnetic induction (MAT-MI). Phys. Med. Biol. **50**(21) (2005)
3. H. Wen, J. Shah, R.S. Balaban, Hall effect imaging. IEEE Trans. Biomed. Eng. **45**(1), 119–124 (1998)
4. C. Gabriel, S. Gabriel, E. Corthout, The dielectric properties of biological tissues: I. Literature survey. Phys. Med. Biol. **41**(11), 2231 (1996)
5. H. Schwan, K. Foster, RF-field interactions with biological systems: electrical properties and biophysical mechanisms. Proc. IEEE **68**(1), 104–113 (1980)
6. T. Coombs, Z. Hong, Y. Yan, C. Rawlings, The next generation of superconducting permanent magnets: The flux pumping method. IEEE Trans. Appl. Supercond. **19**(3), 2169–2173 (2009)
7. R. McGough, *FOCUS*, Michigan State University (2015)
8. J. Gao, K. Wang, J. Sun, Study on the technology of ultrasonic imaging detection based on phase array. Int. J. Signal Process. Image Process. Pattern Recogn. **6**(5), 71–78 (2013)

9.  A. Castellanos, A. Ramos, A. Gonzalez, N.G. Green, H. Morgan, Electrohydrodynamics and dielectrophoresis in microsystems: scaling laws. J. Phys. D Appl. Phys. **36**(20), 2584–2597 (2003)
10. T. Heida, W. Rutten, E. Marani, Understanding dielectrophoretic trapping of neuronal cells: modelling electric field, electrode-liquid interface and fluid flow. J. Phys. D Appl. Phys. **35**(13), 1592–1602 (2002)

# Chapter 7
# Investigation of AC Losses on Stabilizer-Free and Copper Stabilizer HTS Tapes

## 7.1 Introduction

For Direct Current (DC) systems, in principle superconductors should demonstrate electrically lossless features in most conditions [1]. Chapters 4, 5 and 6 presented the design of superconducting magnets for a LFEIT system. Most of the time, these super-conducting magnets are operating in DC conditions. However, when high current superconducting coils and cables are used in magnet applications (e.g. MRIs and LFEITs), they experience power dissipation as they are subjected to the varying magnetic field during the transient operations. Although no actual alternating currents have been involved, this problem generally goes under the classification of "AC loss", because the dissipation dynamics take place during a field ramp, which are substantially the same as the problems encountered in AC conditions. Furthermore, the DC superconducting magnets for both MRIs and LFEITs suffer various ranges of external AC signal disturbances during their operation. A small AC magnetic field disturbance of 10 mT over 100 kHz can affect the stability of a superconducting magnet. Therefore, it is important to investigate the AC loss attribute of HTS tapes and coils for superconducting magnets, even though these magnets are usually operated in DC conditions.

For Alternating Current (AC) systems, superconductors always suffer AC losses in the presence of AC currents and AC magnetic fields. AC loss is an inevitable and crucial issue in AC superconducting systems, particularly in large AC systems such as AC power transmission systems and large-scale superconducting motors for wind turbines and potential aircraft [1, 2]. AC losses could decrease the overall efficiency and create massive problems for cryogenic systems [1, 2].

Second Generation (2G) High Temperature Superconducting (HTS) tapes are the basic elements for HTS coils and various superconducting power applications [3]. As shown in Fig. 7.1, there are two typical types of manufacturing process for 2G

© The Editor(s) (if applicable) and The Author(s), under exclusive license
to Springer Nature Switzerland AG 2020
B. Shen, *Study of Second Generation High Temperature Superconductors:
Electromagnetic Characteristics and AC Loss Analysis*, Springer Theses,
https://doi.org/10.1007/978-3-030-58058-2_7

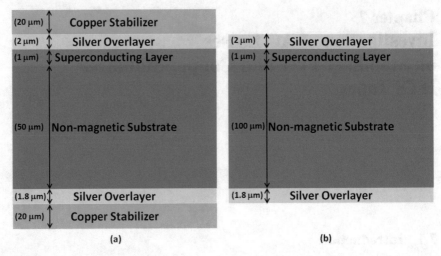

**Fig. 7.1** Cross-section of **a** Surround copper stabilizer (SCS) tape, **b** Stabilizer-free (SF) tape

HTS tapes: Surround Copper Stabilizer (SCS) Tape and Stabilizer-free (SF) Tape [4]. SCS tapes are widely used in high-voltage applications, as the copper stabilizers protect the conductor by bypassing the overcurrents [4]. SF Tapes do not use copper stabilizer, but do have a silver overlayer [4]. SF Tape also has thicker Hastelloy substrates, which are non-magnetic and have high resistivity. Generally, SF Tapes are suitable for power grid protection devices like Superconducting Fault Current Limiters (SFCL) [4]. Therefore, the investigation into AC losses of SCS Tapes and SF Tapes is necessary for research into high-voltage applications and grid protection devices.

The AC loss features of SCS Tape and SF Tapes are different from those researches of tapes with magnetic substrate [5, 6]. Non-magnetic Hastelloy substrates with high resistivity lead to both lower ferromagnetic AC losses and lower eddy current AC losses [7].

Some previous work has been done on Superconducting SFCL experiments and conceptual designs which use the tapes with a non-magnetic substrate [7, 8]. However, according to the best literature search, there has been no research into the comparison of AC losses from SCS Tape and SF Tape with systematic analysis over the wide range of frequency dependence. This Chapter presents the AC loss measurement of a single SCS Tape and an SF Tape using the electrical method, comparing with both real geometry and multi-layer HTS tape simulation using the FEM package of COMSOL Multiphysics. The frequency dependence (up to kHz) of AC losses on SCS Tape and SF Tape are then analysed according to the experimental and simulation results.

## 7.2   Experiment Set-Up

Figures 7.2, 7.3, 7.4, 7.5 and 7.6 show the main components for AC loss measurement using the electrical method.

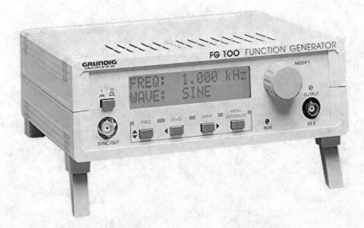

**Fig. 7.2**  Function generator (Digimess® FG100)

**Fig. 7.3**  Power amplifier (Carlsbro Powerline Pro 1200)

**Fig. 7.4**  Lock-in amplifier (signal recovery 7265)

**Fig. 7.5** Step-down transformer

**Fig. 7.6** Compensation coil (fabricated by two co-axis coils)

Figure 7.7 illustrates the schematic of the electrical method to measure the AC losses of an SF Tape (SuperPower SF12100, 12 mm wide) or an SCS Tape (Super-Power SCS12050, 12 mm wide). Their critical currents $I_c$ were measured to both be approximately 300 A with self-fields.

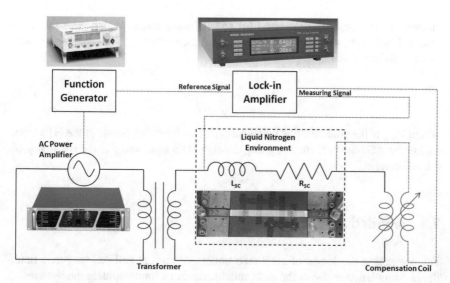

**Fig. 7.7** Experimental schematic of AC loss measurement using the electrical method

The Function Generator (Digimess® FG100) created the identical AC small signal for both the reference input of the Lock-in Amplifier (Signal Recovery 7265) and the Power Amplifier (Carlsbro Powerline Pro 1200). Then, the Power Amplifier generated an AC current in the primary circuit. By using a step-down transformer, the AC current in the secondary circuit was increased 16 times higher than the primary circuit. In the secondary side, the HTS tape was immersed into liquid nitrogen at 77 K, with the measuring technique of "8" glyph potential leads which could effectively reduce the measurement noise [9]. The length between two soldering points for voltage measuring leads was 90 mm.

Both the AC currents in the primary side and secondary side were monitored by a high accuracy Data Acquisition Card linked to NI SignalExpress. Currents were calculated using the voltage across the shunt resistors divided by the resistances. The compensation coil was fabricated by two co-axis coils (1st coil in the high current side, and 2nd coil in the measuring signal side). The secondary side of the compensation coil was made from flexible litz wire (LI14, 0.04 mm²), in order to reduce the eddy-current losses from the compensation system. Obtaining the resistive voltage component (in-phase voltage component) is a key step for AC loss measurement of HTS tapes. Actually, there were two possible methods to get the in-phase voltage component. (a) Extracting a resistive voltage component from the shunt resistor in the circuit as the reference signal, and then using the lock-in amplifier to pick up the in-phase voltage component from the superconducting tape; (b) Using the function generator signal as both the source for AC system and the reference signal for lock-in amplifier, and then using the adjustable compensation coil to compensate the inductive voltage quantity (to get the minimum voltage) in the measuring signal side, which enables the lock-in Amplifier to extract the voltage in-phase with the current of HTS tape. Both method (a) and (b) were tested, which gave the same results. Method (b) was used to carry out all the measurement. The transport AC loss can be calculated as [2]:

$$Q_{ac\_loss} = \frac{V_{rms} \cdot I_{rms}}{f} \tag{7.1}$$

where $V_{rms}$ is the pure resistive voltage across two soldering points of the HTS tape, $I_{rms}$ is the AC transport current going through HTS tape, and $f$ is the frequency of the AC current.

## 7.3  Simulation Method

For computing the losses of HTS tape under AC current and AC magnetic field, the $H$-formulation is one of the most suitable methods for computing the hysteresis, ferromagnetic and eddy-current losses at different layers of the HTS tape [1]. Similar to the modelling of a superconducting magnet in Chaps. 4–6, AC loss calculation uses the $H$-formulation which combines Maxwell Ampere's Law, Faraday's Law, the Constitutive Law, Ohm's Law and the $E$-$J$ power Law [10–15]. The 2D $H$-formulation model using a Cartesian coordinate model by 2 variables $H = [H_x, H_y]$, was solved by COMSOL Multiphysics.

The simulation models were based on the real geometry of SuperPower SCS12050 and SuperPower SF12100 for the COMSOL simulation. SCS12050 has a 1 μm superconducting layer, 20 μm copper stabilizers (upper and lower layer), a 2 μm upper silver overlayer, a 1.8 μm lower silver overlayer, and a 50 μm substrate layer. SF12100 has thicker substrate at 100 μm, but without copper stabilizers. Figure 7.8

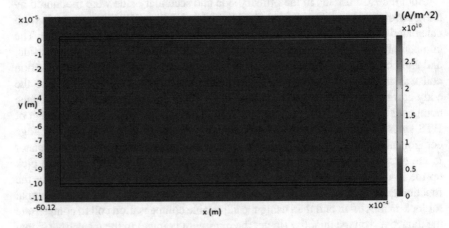

**Fig. 7.8** Example for the 2D cross-section of Stabilizer-free tape (near to the left edge) with a superconducting layer, silver overlayers, and substrate

**Table 7.1** Parameters for the simulation of SuperPower SCS12050 and SF12100 tape

| Parameters | Value |
|---|---|
| Tape width | 12 mm |
| Superconducting layer thickness | 1 μm |
| Upper copper stabilizer thickness (SCS12050 only) | 20 μm |
| Lower copper stabilizer thickness (SCS12050 only) | 20 μm |
| Upper silver overlayer thickness | 2 μm |
| Lower silver overlayer thickness | 1.8 μm |
| Substrate thickness (SF12050) | 50 μm |
| Substrate thickness (SF12100) | 100 μm |
| Silver resistivity (77 K) | $2.7 \times 10^{-9}$ Ω·m |
| Copper resistivity (77 K) | $3.0 \times 10^{-9}$ Ω·m |
| Substrate resistivity (77 K) | $1.25 \times 10^{-6}$ Ω·m |
| Critical current $I_c$ (self-field) | 300 A |
| $\mu_0$ | $4\pi \times 10^{-7}$ H/m |
| $n$ (E-J Power Law factor) | 30 |
| $J_{c0}$ | $2.5 \times 10^{10}$ A/m$^2$ |
| $E_0$ | $10^{-4}$ V/m |

shows an example for the 2D cross-section of SF12100 tape (near to the left edge).

Some relevant parameters are in Table 7.1, e.g. the resistivity of copper is $3.0 \times 10^{-9}$ Ω·m at 77 K, the resistivity of silver is $2.7 \times 10^{-9}$ Ω·m at 77 K [16], and the resistivity of the substrate is $1.25 \times 10^{-6}$ Ω·m [4].

The $B$-dependent critical current model was used for the COMSOL simulation, because $J_c$ can be varied in the presence of parallel and perpendicular magnetic fields [17]:

$$J_c(B) = \frac{J_0}{\left(1 + \sqrt{\frac{k^2 B_{para}^2 + B_{perp}^2}{B_0^2}}\right)} \tag{7.2}$$

where $k = 0.186$ and $B_0 = 0.426$ were used in (7.2), and $J_0$ is the critical current in a self-field at 77 K.

Pointwise or Global constraints from the general PDE Physics enabled transport current were applied to the HTS tape [17]. The integration of current density $J$ over the superconducting domain $\Omega$ was equal to the magnitude of the transport current $I_t$:

$$I_t = \int_{\Omega} J \ d\Omega \qquad (7.3)$$

As the substrates of SuperPower SCS12050 and SF12100 are non-magnetic, the ferromagnetic AC losses of the substrate were ignored. Therefore, for the simulation, the hysteresis losses in the superconducting layer and the eddy-current losses in the copper stabilizer, silver overlayer and substrate are the main factors of total AC losses. The AC loss of the domain was calculated by integrating the power density ($EJ$) over the domain and time [2]:

$$Q = \frac{2}{T} \int_{0.5T}^{T} \int_{\Omega} E \cdot J \ d\Omega dt \qquad (7.4)$$

where $T$ is the period of the cycle and $\Omega$ is the domain of interest.

## 7.4  Results and Discussion

### 7.4.1  Simulation of Eddy-Current AC Losses in Different Layers of SCS Tape

Figure 7.9 presents the simulation of the total eddy-current losses in copper stabilizer (2 layers), silver overlayer (2 layers) and substrate (1 layer) of SCS12050 Tape, with respect to the frequency of the transport current (300 A peak, 100% of $I_c$) which increased from 10 to 1000 Hz. It can be discovered that the eddy-current AC losses of copper stabilizer substrate were approximately 10 times higher than those of the silver overlayer, and over 2 orders of magnitude higher than the eddy-current AC losses from the substrate. The simulation result matched the typical eddy-current equation [1]:

$$P_{\text{eddy}} = \frac{4\mu_0^2}{\pi} \frac{twf^2}{\rho} I_c^2 h(i) \qquad (7.5)$$

where $\rho$ is the resistivity of layer material, $t$ and $w$ and are the thickness and width, and $h(i)$ is a function with normalized operation current.

The simulation results were consistent with Eq. (7.5) that the eddy-current losses are inversely proportional to the resistivity of the different layers, and proportional to the second power of the frequency. The resistivities of copper and silver are comparable when they are at 77 K, and the total thickness of the copper stabilizer is 10 times greater than the thickness of the silver overlayer. This was the reason why the copper stabilizer had one order higher eddy current losses than the silver

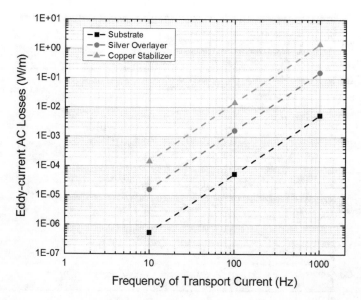

**Fig. 7.9** Simulation of the total eddy-current AC losses from copper stabilizer (2 layers), silver overlayer (2 layers), and substrate (1 layer) of SCS12050 Tape, with the frequency of transport current f increasing from 10, 100 to 1000 Hz, at a peak current of 300 A (100% of $I_c$)

overlayer, and the eddy-current losses from the copper stabilizer dominated the total eddy-current losses of the SCS Tape.

## 7.4.2 Comparison of SF Tape and SCS Tape with Various Frequencies of Transport Current

Figures 7.10, 7.11 and 7.12 shows the comparison of experimental AC losses from SCS Tape and SF Tape, as well as the simulation of hysteresis AC losses in the super-conducting layer, overall eddy-current AC losses in all the layers, and the total losses (superconductor hysteresis losses added to eddy-current losses) from the simulation. Norris's analytical solutions for both the Strip and Ellipse cases are also shown in Figs. 7.10, 7.11 and 7.12. The frequency of the transport current changed from 10, 100 to 1000 Hz. For all three cases, hysteresis AC losses in the superconducting layer from simulation agreed well with Norris's Strip characteristic, which were frequency independent.

In the 10 Hz case (Fig. 7.10), for both the SCS Tape and SF Tape, the simulation results demonstrated that the total eddy-current AC losses were always far lower than the hysteresis AC losses in the superconducting layer. Therefore, the total losses curve clearly indicates that the hysteresis AC losses dominated. The experimental results were between the Norris's curves for Strip and Ellipse, and agreed with the total

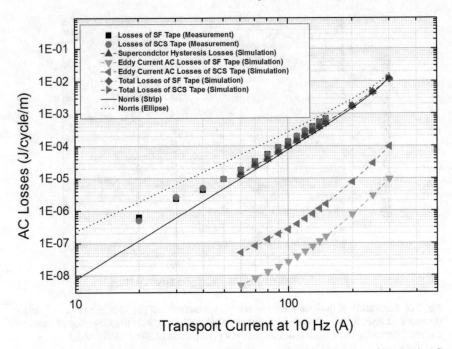

**Fig. 7.10** Measured AC losses in SCS Tape and SF Tape, COMSOL simulation of hysteresis AC losses in the superconducting layer, overall eddy-current AC losses, and total losses of the SCS tape and SF tape, with reference to Norris's analytical solutions (strip and ellipse), at a transport current frequency of 10 Hz

losses curve of the simulation. The measured AC loss data points of SCS Tape and SF Tape were well matched.

For the 100 Hz case (Fig. 7.11), the eddy-current AC losses were similar to the 10 Hz case. The simulation showed that the total eddy-current AC losses were still much lower than the hysteresis AC losses in the superconducting layer for both the SCS Tape and SF Tape cases. The experimental results were consistent with the total loss curves from the simulation with the 100 Hz case. The loss measurements of the SCS Tape were in good agreement with the loss measurements of the SF Tape.

For the 1000 Hz case (Fig. 7.12), the simulation results showed that the eddy-current AC losses of copper stabilizer increased and started to affect the total losses of the SCS Tape, which could be observed in the simulation in that the curve for the total losses of the SCS Tape was slightly above the hysteresis losses of the superconducting layer, and became more obvious with a higher transport current. However, the eddy-current AC losses of the silver overlayer at 1000 Hz were still much smaller than the hysteresis losses of the superconducting layers, which were much less significant. The measured AC losses data points of the SCS Tape were also slightly higher than the measured losses of the SF Tape at a transport current of 1000 Hz, which implies that the eddy-current AC losses of the copper stabilizer slightly affected the total AC losses of the SCS Tape.

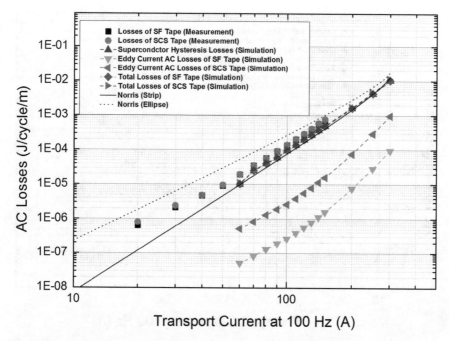

**Fig. 7.11** Measured AC losses in SCS tape and SF tape, COMSOL simulation of hysteresis AC losses in the superconducting layer, overall eddy-current AC losses, and total losses of the SCS tape and SF tape, with reference to Norris's analytical solutions (strip and ellipse), at a transport current frequency of 100 Hz

### 7.4.3   Eddy-Current AC Loss Measurement and Simulation with Frequency at the Kilohertz Level

The eddy-current AC losses of the copper stabilizer could be calculated using the difference in value between the total losses of the SCS Tape and the total losses of the SF Tape, because the only difference between the SCS Tape and the SF Tape was the outer layers of copper stabilizer. This method was valid both for the measurement and simulation. Figure 7.13 illustrates the eddy-current AC losses in the copper stabilizer of SCS12050 Tape determined by the above method, with the AC transport current at 1000 Hz. It reveals that the eddy-current AC losses in the copper stabilizer started to approach Norris's Strip in Joule per cycle value with a frequency of 1000 Hz. From Fig. 7.13, it can be seen that the measurement results matched the simulation results with the same magnitude of transport current.

Due to the limitation of measurement devices, the AC loss measurement with a transport current frequency above 1000 Hz was difficult to achieve, but the estimation of total AC losses from the SF Tape and the SCS Tape could still be carried out using simulation. Figure 7.14 presents the simulation of the SF Tape and the SCS Tape with the AC transport current at 10000 Hz. It can be seen that the eddy-current AC

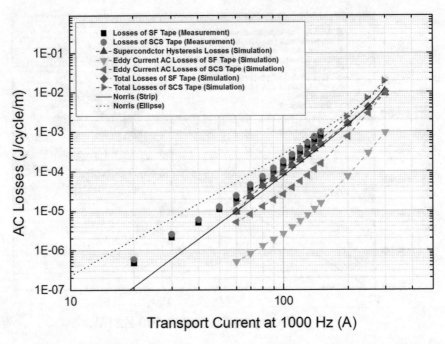

**Fig. 7.12** Measured AC losses in SCS Tape and SF Tape, COMSOL simulation of hysteresis AC losses in the superconducting layer, overall eddy-current AC losses, and total losses of the SCS tape and SF tape, with reference to Norris's analytical solutions (strip and ellipse), at a transport current frequency of 1000 Hz

losses of the silver overlayer at 10000 Hz also slightly affected the total AC losses of the SF Tape. Furthermore, the eddy-current AC losses of the copper stabilizer at 10000 Hz were greater than the hysteresis AC losses in the superconducting layer. From Fig. 7.14, it can be estimated that the eddy-current AC losses of the copper stabilizer occupied a great amount of the total losses of the SCS Tape at a frequency of 10000 Hz.

## 7.5  Summary

The investigation and comparison of AC losses on the SCS Tape and SF Tape have been carried out, which includes AC loss measurement using the electrical method, as well as the real geometry and multi-layer HTS tape simulation using the 2D *H*-formulation by COMSOL Multiphysics. Hysteresis AC losses in the superconducting layer and eddy-current AC losses in the copper stabilizer, the silver overlayer and the substrate have been analysed, because the ferromagnetic AC losses of the substrate were ignored due to the substrates of Superpower SCS12050 and SF12100 being non-magnetic. The results showed that hysteresis AC losses in the superconducting

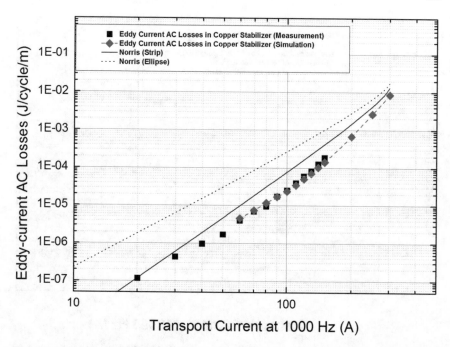

**Fig. 7.13** Comparison between measurement and simulation: eddy-current AC losses in the copper stabilizer in the SCS tape, with reference to Norris's analytical solutions (strip and ellipse), at a transport current frequency of 1000 Hz

layer are frequency independent. The eddy-current AC losses (in Watts) were proportional to the second power of the frequency. The experimental and simulation results revealed that the eddy-current AC losses had almost no effect on the total AC losses for both the SCS Tape and the SF Tape with transport current frequencies of 10 and 100 Hz. However, for a transport current frequency of 1000 Hz, the experiment and simulation showed that eddy-current AC losses started to affect the total losses for the SCS Tape, but that the SF Tape still obtained negligible eddy-current AC losses when the transport current frequency reached 1000 Hz. The eddy-current AC losses in the copper stabilizer were determined using the difference in value between the total losses of the SCS Tape and the total losses of the SF Tape, which was valid for both the measurement and simulation, and the results showed that the measurement matched the simulation. The estimation of AC losses with a frequency of 10000 Hz was also carried out using COMSOL simulation. It could be deduced that the eddy-current AC losses in the copper stabilizer should also be taken into consideration for high frequency applications over the kilohertz level.

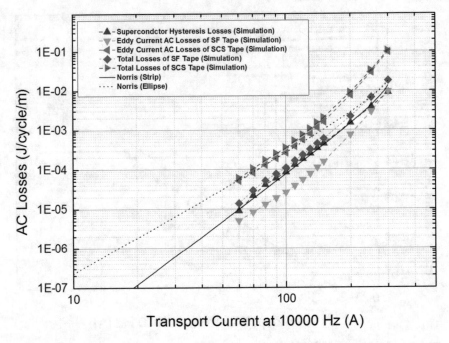

**Fig. 7.14** Simulation of hysteresis AC losses in the superconducting layer, overall eddy-current AC losses, and total losses of the SCS tape and SF tape, with reference to Norris's analytical solutions (strip and ellipse), at a transport current frequency of 10000 Hz

# References

1. F. Grilli, E. Pardo, A. Stenvall, D.N. Nguyen, W. Yuan, F. Gömöry, Computation of losses in HTS under the action of varying magnetic fields and currents. IEEE Trans. Appl. Supercond. **24**(1), 78–110 (2014)
2. B. Shen, J. Li, J. Geng, L. Fu, X. Zhang, H. Zhang, C. Li, F. Grilli, T. A. Coombs, Investigation of AC losses in horizontally parallel HTS tapes. Supercond. Sci. Technol. **30**(7), Art. no. 075006 (2017)
3. B. Shen, L. Fu, J. Geng, X. Zhang, H. Zhang, Q. Dong, C. Li, J. Li, T. Coombs, Design and simulation of superconducting Lorentz force electrical impedance tomography (LFEIT). Phys. C Supercond. **524**, 5–12 (2016)
4. SuperPower, *SuperPower® 2G HTS Wire Specifications*. Schenectady, NY (2014)
5. Y. Mawatari, Magnetic field distributions around superconducting strips on ferromagnetic substrates. Phys. Rev. B **77**(10) (2008)
6. S. Li, D.-X. Chen, J. Fang, Transport ac losses of a second-generation HTS tape with a ferromagnetic substrate and conducting stabilizer. Supercond. Sci. Technol. **28**(12), Art. no. 125011 (2015)
7. J.B. Na, D.K. Park, S.E. Yang, Y.J. Kim, K.S. Chang, H. Kang, T.K. Ko, Experimental analysis of bifilar pancake type fault current limiting coil using stabilizer-free coated conductor. IEEE Trans. Appl. Supercond. **19**(3), 1797–1800 (2009)
8. D.K. Park, S.E. Yang, Y.J. Kim, K.S. Chang, T.K. Ko, M.C. Ahn, Y.S. Yoon, Experimental and numerical analysis of high resistive coated conductor for conceptual design of fault current limiter. Cryogenics **49**(6), 249–253 (2009)

9. Z. Wu, Y. Xue, J. Fang, L. Yin, D. Chen, The influence of the YBCO tape arrangement and gap between the two tapes on AC loss. IEEE Trans. Appl. Supercond. **26**(7) (2016)

10. B. Shen, F. Grilli, T. Coombs, Overview of H-formulation: a versatile tool for modelling electromagnetics in high-temperature superconductor applications. IEEE Access **8**, 100403–100414 (2020)

11. B. Shen, T. Coombs, F. Grilli, AC loss analysis on HTS crossconductor (CroCo) cables for power transmission, in 2018 IEEE international conference on applied superconductivity and electromagnetic devices (ASEMD), Tianjin, China (2018)

12. B. Shen, T. Coombs, F. Grilli, Investigation of AC loss in HTS cross-conductor cables for electrical power transmission. IEEE Trans. Appl. Supercond. **29**(2), Art. no. 5900205 (2019)

13. B. Shen, F. Grilli, T. Coombs, Review of the AC loss computation for HTS using H formulation. Supercond. Sci. Technol. **33**(3), Art. no. 033002 (2020)

14. B. Shen, C. Li, J. Geng, Q. Dong, J. Ma, J. Gawith, K. Zhang, Z. Li, J. Chen, W. Zhou, Power dissipation in the HTS coated conductor tapes and coils under the action of different oscillating currents and fields. IEEE Trans. Appl. Supercond. **29**(5), Art. No. 8201105 (2019)

15. B. Shen, C. Li, J. Geng, Q. Dong, J. Ma, J. Gawith, K. Zhang, J. Yang, X. Li, Z. Huang, T. Coombs, Investigation on power dissipation in the saturated iron-core superconducting fault current limiter. IEEE Trans. Appl. Supercond. **29**(2), Art. no. 5600305 (2019)

16. R.A. Matula, Electrical resistivity of copper, gold, palladium, and silver. J. Phys. Chem. Ref. Data **8**(4), 1147–1298 (1979)

17. B. Shen, L. Fu, J. Geng, H. Zhang, X. Zhang, Z. Zhong, Z. Huang, T. Coombs, Design of a superconducting magnet for lorentz force electrical impedance tomography. IEEE Trans. Appl. Supercond. **26**(3), Art. no. 4400205 (2016)

# Chapter 8
# Study on Power Dissipation in HTS Coils

## 8.1 Introduction

High Temperature Superconductor (HTS) based cables and coils possess the advantage of carrying high electrical current density [1, 2]. HTS coils are used for superconducting power applications, such as superconducting fault current limiters [3, 4] and superconducting transformers [5, 6], as they are able to conduct a higher current but with much less loss than normal metal conductors when they are operating in the superconducting state. HTS coils can greatly increase the magnetic flux density in superconducting electrical machines and magnets, which leads to the improvement of overall efficiency and reduction in weight and size [7, 8], for example, the superconducting magnet for LFEIT in Chaps. 4–6. However, when HTS coils are operating with alternating current (AC), or in the presence of a time varying magnetic field, they still sustain AC losses [9].

The circular HTS coil is one of the most common topologies used in many superconducting applications, such as superconducting transformers, magnetic resonance imaging (MRI) [10], and LFEIT systems as mentioned in Chaps. 4–6. There are several advantages to circular HTS coils. A circular HTS coil is able to generate a uniform magnetic field in the centre region. The round topology of a circular HTS coil offers better mechanical torsion when closely packed.

In the literature, there is some work on the AC loss measurement and simulation of HTS coils. Amemiya et al. have presented the AC loss from HTS tapes under external DC/AC magnetic fields [11]. Ciszek et al. analysed the angular dependence of AC transport losses in HTS tape on external DC magnetic fields [12]. Chiba et al. did research on the angular dependence of the AC loss from HTS stacks [13]. Nguyen et al. carried out AC loss measurement with a phase difference between the current and the applied magnetic field [14]. However, there is no systematic study on AC loss from HTS coils which covers all these aspects mentioned above: current, field,

B. Shen, *Study of Second Generation High Temperature Superconductors:*
*Electromagnetic Characteristics and AC Loss Analysis*, Springer Theses,
https://doi.org/10.1007/978-3-030-58058-2_8

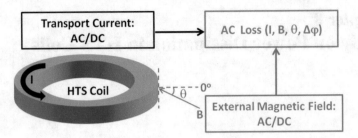

**Fig. 8.1** Factors affecting AC loss in an HTS coil: current, field, angle, and phase difference

angle, and phase difference, as shown in Fig. 8.1 analysed by both experiment and simulation. This chapter presents a comprehensive study on the AC losses in a circular HTS coil, as well as some novel results and analysis. In order to validate some of the results, the experiment was set up to measure the AC loss from a $2 \times 18$ circular double pancake coil using the electrical method. A 2D axisymmetric $H$-formulation model was also built using the FEM package in COMSOL Multiphysics, which was used as the primary tool for this investigation.

As shown in Fig. 8.1, there are several scenarios that can cause AC losses in an HTS coil: (1) AC transport current and DC magnetic field, (2) DC transport current and AC magnetic field, and (3) AC transport current and AC magnetic field. Furthermore, the HTS coil under the magnetic field with different orientation angle $\theta$ has been studied for each case. For the scenario of AC transport current and AC magnetic field, the relative phase difference between the AC current and AC field has been analysed. To summarise, a current/field/angle/phase dependent AC loss $(I, B, \theta, \Delta\varphi)$ study of a circular HTS coil was carried out by both experiment and simulation.

For scenario (1), both the experiment and simulation have been carried out, and very good agreement was seen in terms of AC loss magnitude, tendency, and angular dependence. Therefore, it is convincing that the coil model is capable of producing convincing results for scenarios (2) and (3). Another factor is the increasing complexity of the measurement for the following scenarios. Particularly for scenario (3), both the transport current loss and magnetisation loss require to be measured using separate experimental methods, and if along with the phase difference between the AC current and AC field, the measurement complexity will even further increase and affect the measurement accuracy. Therefore, it is necessary to establish a powerful numerical model for the coil which can efficiently calculate the AC loss under various complex conditions.

## 8.2 Experiment Set-up

### 8.2.1 Fabrication of the Circular HTS Coil

The circular HTS coil used in this experiment was fabricated from 6 mm wide SuperPower SCS6050 tape (manufactured around 2012). The critical current $I_c$ of a single tape in its self–field was measured to be approximately 115 A. The total length of the tape for winding the coil was 6.3 m, whose surface was insulated by low temperature KAPTON tapes. As shown in Fig. 8.2a, the configuration of the coil was a double circular pancake, with $2 \times 18$ turns. The critical current $I_c$ of the coil in its self–field was measured to be approximately 72 A. More details of the circular HTS coil are in Table 8.1.

### 8.2.2 AC Loss Measurement

Figure 8.2a presents the schematic for the measurement of the AC losses in a circular HTS coil using the electrical method. The function generator (Digimess® FG100) produced time-varying sinusoidal signals as the reference input of the lock-in amplifier (Signal Recovery 7265), and this AC signal was also amplified by a power source (Carlsbro Powerline Pro 1200) in the primary circuit side. In the secondary circuit, the AC current was raised 16 times by using a step-down transformer. As shown in Fig. 8.2b, the HTS coil was fixed by a coil holder, and immersed into a liquid nitrogen environment at 77 K. The HTS coil was located in the presence of uniform magnetic field generated by the iron magnet (shown in Fig. 8.2c). The AC current magnitude in the secondary side was obtained using the voltage across the shunt resistor divided by the resistance, and monitored by a high accuracy data acquisition card linked to the PC with the software NI SignalExpress.

The adjustable compensation coil was to compensate for the inductive component from the HTS coils. The lock-in amplifier was used to extract the voltage in-phase with the current. The transport AC loss can be calculated as [5]:

$$Q_{ac\_loss} = \frac{I_{rms} \cdot V_{rms}}{f} \tag{8.1}$$

where $I_{rms}$ is the AC transport current flowing through the HTS tape, $V_{rms}$ is the in–phase voltage with current $I_{rms}$, and $f$ is the frequency of the AC current.

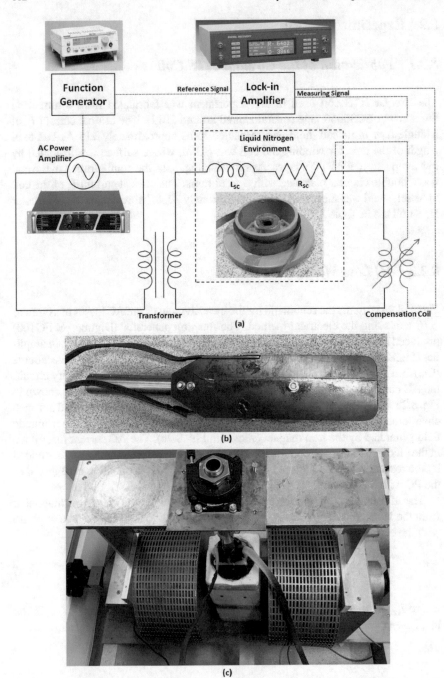

Fig. 8.2 **a** Experimental schematic of AC loss measurements using the electrical method, **b** Coil holder, **c** Electrometric magnet for field dependence analysis

**Table 8.1** Parameters for the circular HTS coil made by SuperPower SCS6050 (2012)

| Parameters | Value |
|---|---|
| Tape width | 6 mm |
| Superconducting layer thickness | 1 μm |
| Tape total thickness | 100 μm |
| KAPTON tape thickness | 100 μm |
| Coil inner diameter | 5 cm |
| Coil total length | 6.3 m |
| Turn number | $2 \times 18$ |
| Tape Self-field $I_c$ at 77 K | 115 A |
| Coil Self-field $I_c$ at 77 K | 72 A |

## 8.3 Simulation Method

### 8.3.1 2D Axisymmetric H-Formulation

The 2D axisymmetric $H$-formulation has been used as the suitable FEM method for modelling the circular HTS coil. An example of the double circular pancake coil using 2D axisymmetric $H$–formulation is presented in Fig. 8.3: (a) 3D and (b) 2D magnetic flux density of a circular double pancake coil with AC current of 60 A (peak point); (c) normalised current density ratio ($J/J_c$) of a circular double pancake coil with an AC current of 60 A (peak point) where the cross-sections of the tape have been zoomed 100 times for better visualisation.

A 2D axisymmetric model of the coil matches the experimental situation only in the case of self-field or when the external magnetic field is applied parallel to the axis of the coil (axis z in Fig. 8.3a, $\theta = 90°$ in Fig. 8.1). In the other situations with external magnetic field, the axial symmetry is broken and a 3D model would be necessary. However, a full 3D model of the coil is computationally too demanding. For this reason, we kept the simplifying assumption of a 2D axisymmetric model, because with the exception $\theta = 90°$, for which the 2D axisymmetric model is appropriate. It implies that there is always a magnetic field component perpendicular to the tape for the whole length of the coil. This represents a kind of worst case scenario and provides a reasonable upper limit for the losses. As presented in Chap. 3, the general $H$-formulation consists of Ohm's Law, Ampere's Law, Faraday's Law, the Constitutive Law, and the $E$-$J$ power Law. The general form of partial differential equation (PDE) for variables $H$ to be computed by COMSOL Multiphysics is:

$$\frac{\partial(\mu_0\mu_r \boldsymbol{H})}{\partial t} + \nabla \times (\rho\nabla \times \boldsymbol{H}) = 0 \tag{8.2}$$

**Fig. 8.3** **a** 3D and **b** 2D
magnetic flux density of a
circular double pancake coil
with AC current of 60 A
(peak point); **c** normalised
current density ratio (J/Jc) of
a circular double pancake
coil with AC current of 60 A
(peak point), where the
cross-sections of tape have
been zoomed for 100 times
for better visualisation

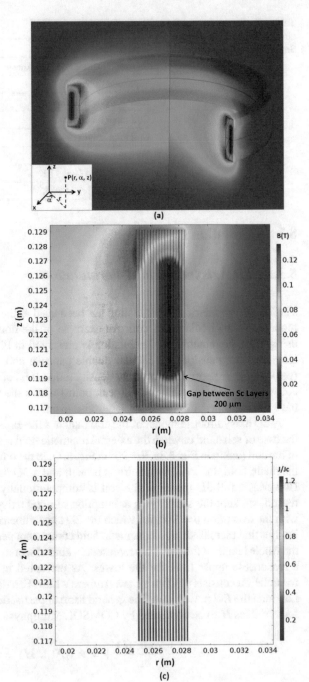

For a 2D axisymmetric $\mathbf{H}$-formulation, the cylindrical coordinates $(r, \theta, z)$ were used, and the governing equations should be modified as:

$$-\frac{\partial E_\theta}{\partial z} = -\frac{\partial(\mu_0 \mu_r H_r)}{\partial t} \tag{8.3}$$

$$\frac{E_\theta}{r} + \frac{\partial E_\theta}{\partial r} = -\frac{\partial(\mu_0 \mu_r H_z)}{\partial t} \tag{8.4}$$

### 8.3.2 AC Loss Calculation

In the FEM model, the real dimension of the SuperPower SCS6050 tape was used, with a superconducting layer 1 μm thick. The geometry of modelling a 2 × 18 turns double circular pancake coil was exactly same as the real experimental coil described above, in order to achieve better consistency. The E-J Power Law factor $n$ used for modelling was 25. This is a moderate value when a single tape is in the presence of DC magnetic field between 0 and 500 mT, and according to measurements, the $n$ variation of the SCS6050 tape is not significant within this field range. The most relevant simulation parameters are listed in Table 8.2. For the modelling of HTS coil by COMSOL, an anisotropic $\mathbf{B}$-dependent critical current model was implanted [15]:

$$J_c(B) = \frac{J_{c0}}{\left(1 + \dfrac{\sqrt{(kB_{\mathrm{para}})^2 + B_{perp}^2}}{B_c}\right)^b} \tag{8.5}$$

where $J_{c0} = 2.1 \times 10^{10}$ A/m$^2$, $k = 0.25$, $B_c = 0.3$, and $b = 0.6$. The critical current of tape and coil were calculated with the same $\mathbf{H}$-formulation model using a slow

| **Table 8.2** Parameters for the modelling of a circular HTS coil | Parameters | Value |
|---|---|---|
| | Tape width | 6 mm |
| | Superconducting layer thickness | 1 μm |
| | $\mu_0$ | $4\pi \times 10^{-7}$ H/m |
| | $n$ (E-J Power Law factor) | 25 |
| | $J_{c0}$ | $2.1 \times 10^{10}$ A/m$^2$ |
| | $E_0$ | $10^{-4}$ V/m |
| | $B_c$ | 35 mT |
| | $k$ | 0.25 |
| | $b$ | 0.6 |
| | $f$ | 50 Hz |

current ramp. The calculated critical current of the single tape in self-field was 114.5 A, and the critical current of the double pancake coil was 71.6 A. These two critical current values are very close to the experimental results (single tape $I_c = 115$ A, and coil $I_c = 72$ A) mentioned above, which proves a good consistency between modelling and experiment.

There were no ferromagnetic losses as the SuperPower SCS6050 tape uses a non–magnetic substrate. The simulations were carried out using a relatively low power frequency AC current at 50 Hz, and the small amount of eddy-current AC losses in the metal layers and the substrate were ignored.

In the model, the transport current was injected into the HTS tapes using the Global constraint from general PDE Physics. The value of the transport current $I_t$ was calculated by the integration of the current density $J$ on the superconducting domain $\Omega$:

$$I_t = \int_\Omega J \, d\Omega \tag{8.6}$$

The AC loss of the domain was calculated using the power density $(E \cdot J)$ integration [16]:

$$Q = \frac{2}{T} \int_{0.5T}^{T} \int_\Omega 2\pi r E \cdot J \, d\Omega dt \tag{8.7}$$

where T is the period of cycle and $\Omega$ is the domain of interest. The AC loss per unit length can be obtained by dividing the value calculated from (8.7) by the coil's length.

## 8.4   Results: AC Transport Current and External DC Magnetic Field

### 8.4.1   Basic Frequency Dependence Test of AC Loss for the HTS Tape and Coil

To start with the simplest situation, AC loss measurement and simulation of a single tape, the AC loss measurement was carried out using frequencies of 50, 100, and 200 Hz, and a transport current from 20 to 100 A. The FEM simulation of a single tape was also performed with an AC transport current from 20 to 100 A at 50 Hz. In Fig. 8.4, the measurement results show that the AC losses of a single tape were frequency independent within the range of 50–200 Hz. As shown in Fig. 8.4, both the AC losses from the experiment and simulation were within the range of the Norris

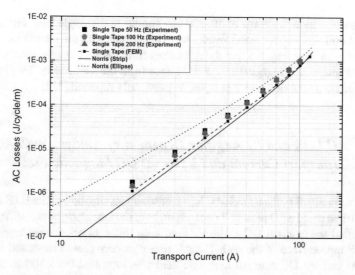

**Fig. 8.4** AC loss measurement and simulation of a single tape, with reference to the Norris strip and Norris ellipse

strip and Norris ellipse, and the experimental results agreed well with the simulation results.

Similarly, the AC losses of the circular double pancake HTS coil were measured with frequencies of 50, 100 and 200 Hz, and a transport current from 20 to 70 A. The FEM calculation of that coil tape was also performed using the same AC transport current at 50 Hz. As presented in Fig. 8.5, the AC losses of the coil were frequency

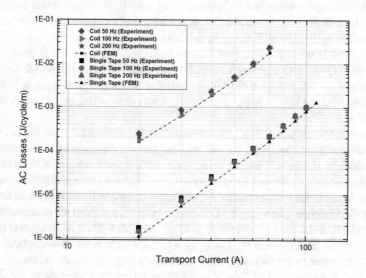

**Fig. 8.5** Comparison of HTS coil and single tape: AC loss measurement and simulation

independent within the range of 50–200 Hz, similar to the single tape. For both the tape and coil, the experimental results were consistent with the simulation results. It can be seen from Fig. 8.5 that the AC losses from the coil were approximately 2 orders of magnitude higher than the AC losses from the single tape. This is due to the magnetic interaction between tapes, as previously reported [17].

### 8.4.2    AC Loss and Its Angular Dependency Analysis with AC Transport Current and External DC Magnetic Field

As shown in Fig. 8.6, the overall AC loss measurement situation of the HTS coil can be described using a 3D plot, as it depends on both the amplitude of the AC transport current and the external DC magnetic field with different orientation angles, $\theta$ to the HTS tape surface of the coil. The AC transport current was increased from 20 to 60 A, and the DC external magnetic field was increased from 100 to 300 mT. The angle $\theta$ refers to the external magnetic field with different orientation to the HTS tape surface of the coil (e.g. as shown in Fig. 8.1, $\theta$ changing from $-90°$ to $+90°$ means that the magnetic field orientation to the HTS tape surface of the coil changes from parallel to perpendicular, then to parallel again). Figure 8.6 shows that the biggest losses occurred when the angle $\theta$ was $0°$, for all the amplitudes of external DC magnetic field (100, 200 and 300 mT). For lower transport currents, the difference in AC loss between each external DC magnetic field was larger, while for higher transport currents the loss difference of each DC external magnetic field became smaller.

Figure 8.7 presents the AC loss measurement and simulation of the HTS coil, with AC transport current and DC external magnetic field perpendicular to the HTS tape surface of the coil. From Fig. 8.7, it can be seen from both the experiment and the simulation that the difference in AC loss with each set of external DC fields became smaller when the transport current increased, e.g. from the experiment the loss ratio of $Q_{300mT}/Q_{0mT}$ was 5.1 with 20 A AC transport current, which decreased to 1.2 when the AC transport current increased to 60 A. this phenomenon is similar to that reported in [11]. Overall, the same trend was observed in both experiment and simulation.

Figure 8.8 illustrates the AC loss measurement and simulation of the HTS coil, with increasing AC transport current and angular dependence of the external DC magnetic field 100 of mT, while Fig. 8.9 presents the same content with an external DC magnetic field of 300 mT for comparison. For both the 100 and 300 mT cases, the angular dependence with a smaller AC transport current was more obvious than the angular dependence with a larger AC transport current, e.g. from the experiment with a 300 mT DC field, the loss ratio of $Q_{0°}/Q_{90°}$ was 7.0 with an AC transport current of 20 A, while the loss ratio of $Q_{0°}/Q_{90°}$ was 1.29 with an AC transport current of 60 A. This is due to the fact that, when a relatively large transport AC current flows in the coil, the external DC field is less influential on the reduction of the critical

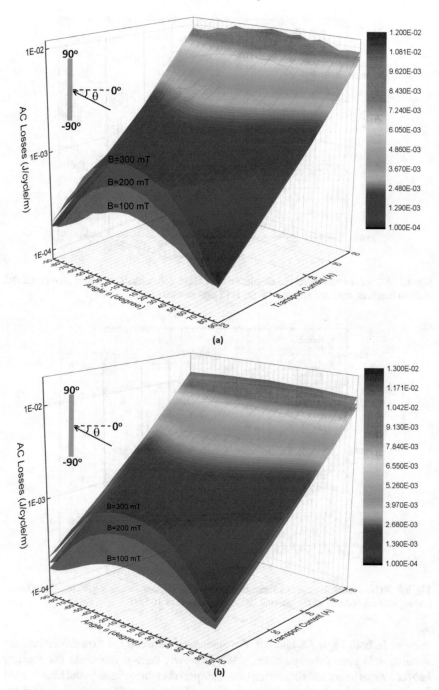

**Fig. 8.6** **a** AC loss measurement and **b** AC loss simulation of an HTS coil using a 3D plot, with the relationship of the AC transport current amplitude and the DC external magnetic field with different orientation angles, $\theta$, to the HTS tape surface of the coil

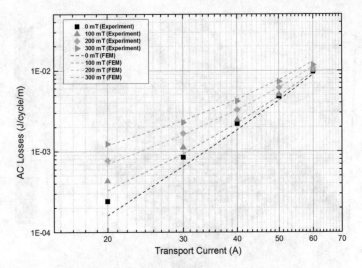

**Fig. 8.7**  AC loss measurement and simulation of the HTS coil, with AC transport current and DC external magnetic field perpendicular to the HTS tape surface of coil

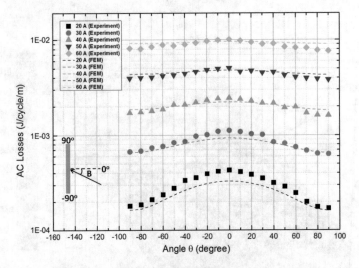

**Fig. 8.8**  AC loss measurement and simulation of the HTS coil, with increasing AC transport current and angular dependence of the external DC magnetic field of 100 mT

current. In both Figs. 8.8 and 8.9, the agreement between the experiment and the simulation is good (average error 3.2%). However, there was a slight discrepancy between experiment and simulation (low transport current in Fig. 8.8, and high current in Fig. 8.9), which is probably due to local effects (e.g. uniformity of $J_c$ near the edges, alignment of the tapes) that are not considered by the model.

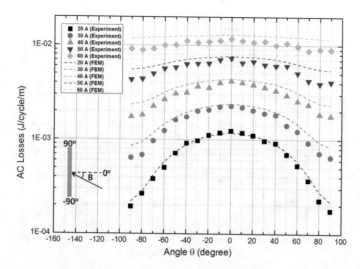

**Fig. 8.9** AC loss measurement and simulation of the HTS coil, with increasing AC transport current and angular dependence of the external DC magnetic field of 300 mT

Figure 8.10 shows the AC loss measurement and simulation of the HTS coil, with the same AC transport current of 30 A and angular dependence of the external DC magnetic field increasing from 100, 200 to 300 mT, while Fig. 8.11 presents the same content with an AC transport current of 60 A for comparison. As shown in Fig. 8.10, for a transport current of 30 A, the AC loss was more angular dependent

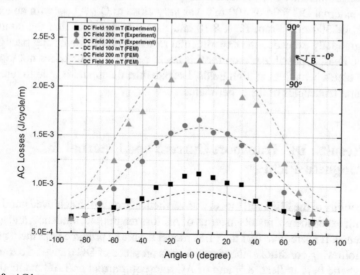

**Fig. 8.10** AC loss measurement and simulation of the HTS coil, with the same AC transport current of 30 A, and angular dependence of the external DC magnetic field increasing from 100, 200 to 300 mT

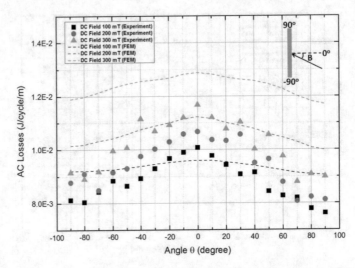

**Fig. 8.11** AC loss measurement and simulation of the HTS coil, with the same AC transport current of 60 A and angular dependence of the external DC magnetic field increasing from 100, 200 to 300 mT

with a stronger external DC magnetic field, e.g. from the experiment, the loss ratio of $Q_{0°}/Q_{90°}$ was 1.8 with an external DC field of 100 mT, whilst $Q_{0°}/Q_{90°}$ was 3.5 with an external DC field of 300 mT. By contrast, as shown in Fig. 8.11, for a higher transport current of 60 A, the AC loss was no longer angular dependent with stronger intensity of external DC magnetic field, e.g. the loss ratio of $Q_{0°}/Q_{90°}$ with an external DC field of 100 mT was very close to $Q_{0°}/Q_{90°}$ with an external DC field of 300 mT. From Figs. 8.10 and 8.11 one can note that the simulation results gradually exceeded the experimental results with the increasing background magnetic field. A possible reason for this could be that Eq. (8.5) does not represent well the tape's behaviour at high fields. Despite this, the model is able to reproduce the general tendencies of the experiments.

## 8.5   Results: DC Transport Current and External AC Magnetic Field

The capability of the FEM coil model was verified in Sect. 8.4. It achieved good agreement with experimental results in terms of AC loss magnitude, tendency, and angular dependence. It is credible that the FEM model is able to produce convincing results for the following conditions: the simultaneous presence of DC transport current and AC magnetic field in Sect. 8.5, and of AC transport current and AC magnetic field later in Sect. 8.6.

### 8.5.1   Basic Test of AC Loss for an HTS Tape and Coil with a DC Transport Current and External AC Magnetic Field

A single SCS6050 tape was simulated in the presence of an AC magnetic field to test the basic consistency. The AC loss of a single SCS6050 tape was calculated under an AC magnetic field with three different frequencies: 50, 100, and 200 Hz. As shown in Fig. 8.12, the AC loss of the single tape was frequency independent, and the loss trend well matched the Brandt curve.

As presented in Fig. 8.13, the AC loss simulation of the coil with different DC transport currents (0, 20, and 40 A) and an increasing external AC magnetic field was carried out, with reference to the AC loss from a single tape in the presence of the same external AC magnetic field. Figure 8.13 shows that, in the coil, the average AC loss per tape was lower than the AC loss of the single tape, and even the AC loss per cross-sections (2 × 18 tapes) was still lower than the AC loss of the single tape with an external AC magnetic field below 200 mT. This is because of the shielding effect from each turn of the HTS coil. However, this effect became weaker with an increasing external AC magnetic field. The increasing rate of loss in the coil was faster than the increasing rate of loss in the single tape, and the average coil AC loss per tape had the trend to surpass the AC loss of the single tape.

In principle, in the presence of the same AC magnetic field, the magnetisation loss of a superconducting tape with an increasing DC transport current should not change (if the tape is not fully penetrated). However, it should be considered that the DC transport current within the HTS coil generated the self–field which could

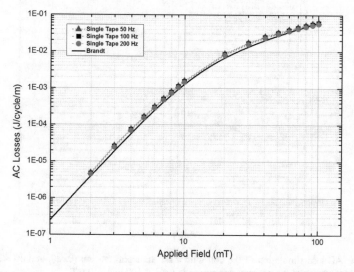

**Fig. 8.12** AC loss simulation of the single tape in the presence of an external AC magnetic field with frequencies of 50, 100 and 200 Hz, with reference to the Brandt curve

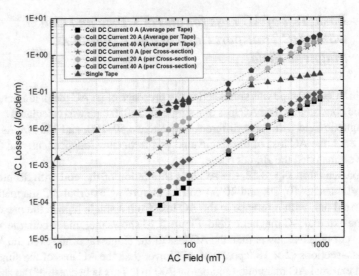

**Fig. 8.13** AC loss simulation of the coil with different DC transport currents (0, 20, and 40 A) and an increasing external AC magnetic field, with reference to the AC loss from a single tape

decrease the critical current of the coil, thus increasing the AC loss of the coil. That phenomenon occurred in Figs. 8.13 and 8.14, and it can be observed that the difference of coil AC loss between the DC transport currents of 20 and 40 A was greater than that of the difference between the DC transport currents of 0 and 20 A,

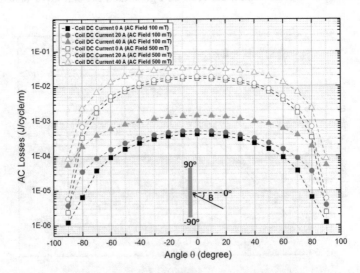

**Fig. 8.14** AC loss simulation of the coil with a DC transport current (0, 20, and 40 A) and the angular dependence of the external AC magnetic field at 100 and 500 mT

which could be because the critical current reduction between DC currents of 20 and 40 A is more significant than between DC currents between 0 and 20 A.

### 8.5.2   AC Loss and Its Angular Dependency Analysis with DC Transport Current and External AC Magnetic Field

Figure 8.14 shows the AC loss simulation of the coil with a DC transport current (0, 20, and 40 A) and the angular dependence of the external AC magnetic field at 100 and 500 mT. The angular dependency with a lower DC transport current was stronger than that with a higher transport current. For example, from the simulation with a 100 mT AC field, the loss ratio of $Q_{0°}/Q_{90°}$ was 342.5 with an DC transport current of 0 A, whilst the loss ratio of $Q_{0°}/Q_{90°}$ was 26.4 with an DC transport current of 40 A. From the comparison of the two cases of AC fields of 100 and 500 mT, it can be seen that the angular dependence was more apparent with the larger external AC field, e.g. from the simulation with a 20 A DC transport current, the loss ratio of $Q_{0°}/Q_{90°}$ was 136.6 with an AC magnetic field 100 mT, while the loss ratio of $Q_{0°}/Q_{90°}$ was 3245.7 with an AC magnetic field of 500 mT.

These two phenomena above are similar to Scenario (1): AC transport current & DC external magnetic field. However, the effect of increasing the DC transport current with a given (fixed) AC external field is different from increasing the AC transport current with a given (fixed) DC external field. This is because the AC loss was substantially caused by two different sources, the AC transport current and the AC external magnetic field, which have different magnetic field penetration patterns.

Figure 8.15 depicts the AC loss of the coil for two sets of DC transport currents (30 and 60 A) and the angular dependence of increasing the external AC magnetic field (100, 200, and 300 mT). For the same DC transport current, the angular dependence was more evident with an increasing external AC magnetic field. The angular dependence did not change too much between the two sets of transport currents, 30 and 60 A, which can be seen from the fact that the loss curves of the 30 and 60 A transport currents were almost parallel. This phenomenon is different from scenario (1), where the angular dependence with a higher transport current was much smaller than that with a lower transport current.

## 8.6   Results: AC Transport Current and External AC Magnetic Field

In this section, the coil AC loss was calculated under the action of an AC transport current and an external AC magnetic field, and its angular dependency was compared to the experimental and simulation results under the action of an AC transport current and an external DC magnetic field. The effect of phase difference was investigated:

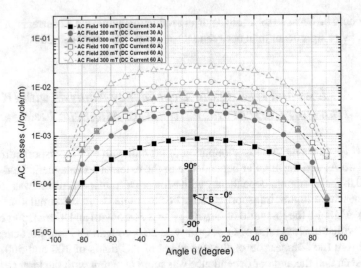

**Fig. 8.15** AC loss simulation of the coil with two sets of the same DC transport current (30 and 60 A) and the angular dependence of increasing the external AC magnetic field (100, 200, and 300 mT)

$\Delta\varphi$ between the AC transport current and the external AC magnetic field on the total AC loss of the coil.

### 8.6.1   AC Losses in an HTS Coil with an AC Transport Current and an External AC Magnetic Field Perpendicular to the HTS Tape Surface of the Coil

Figure 8.16 presents the AC loss simulation of an HTS coil with AC transport current and an external AC magnetic field perpendicular to the HTS tape surface of the coil. The AC loss increasing rate of 0 A transport current started with a steep slope, but the rate decreased with the increasing external AC magnetic field. By contrast, the other three AC loss curves for the AC transport currents 20, 40 and 60 A, started with mild slopes but later gradually increased with the increasing external AC magnetic field. Eventually, these four loss curves for AC transport currents 0, 20, 40 and 60 A moved closer when the external AC magnetic field approached a high value. This tendency is consistent with the work in literature [11].

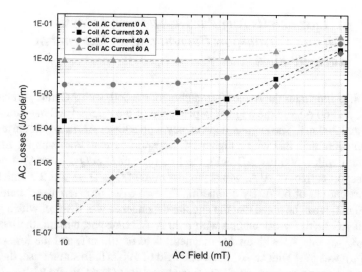

**Fig. 8.16** AC loss simulation of the HTS coil, with an AC transport current and an external AC magnetic field perpendicular to the HTS tape surface of the coil

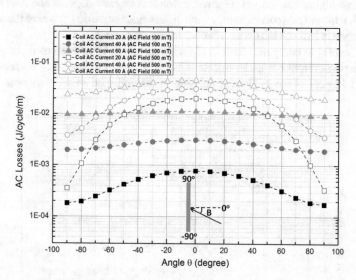

**Fig. 8.17** AC loss simulation of the coil, with AC transport currents (20, 40, and 60 A) and the angular dependence of the external AC magnetic field (100 and 500 mT)

## 8.6.2  AC Loss and Its Angular Dependency Analysis with an AC Transport Current and an External AC Magnetic Field

Figure 8.17 illustrates the AC loss simulation of the coil, with AC transport currents (20, 40, and 60 A) and the angular dependence of the external AC magnetic fields of 100 and 500 mT. The angular dependency with a lower AC transport current was more apparent than that with a higher DC transport current. For example, from the simulation with a 500 mT AC field, the loss ratio of $Q\_0°/Q\_90°$ was 59.4 with an AC transport current of 20 A, whilst the loss ratio of $Q\_0°/Q\_90°$ was 2.3 with an AC transport current of 60 A. By comparing the two cases of external AC fields, 100 and 500 mT, it can be noted that the angular dependence was stronger with a greater external AC field. By calculation with a 20 A AC transport current, the loss ratio of $Q\_0°/Q\_90°$ was 4.5 with an AC magnetic field of 100 mT, and the loss ratio of $Q\_0°/Q\_90°$ was 59.4 with an AC magnetic field of 500 mT. To summarise, these two phenomena above occurred in all the three scenarios: (1) AC $I_t$ & DC $B_{ext}$, (2) DC $I_t$ & AC $B_{ext}$, and (3) AC $I_t$ & AC $B_{ext}$. However, for scenarios (1) and (3), the AC loss with a higher transport current had much less angular dependence than the AC loss with a lower transport current, which differs from scenario (2) where the angular dependence is still obvious with a higher transport current.

It is helpful to compare scenario (3) with scenario (1), as both of them used an AC transport current but under different external fields (AC and DC, respectively). Figure 8.18 depicts the AC loss simulation of the coil with an AC transport current

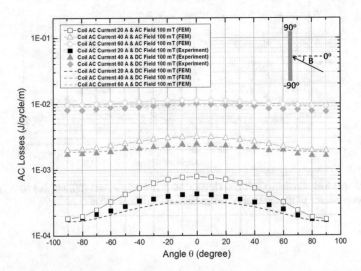

**Fig. 8.18** AC loss simulation of the coil with AC transport currents (20, 40, and 60 A) and the angular dependence of an external AC magnetic field of 100 mT, and its comparison to the AC loss experiment and simulation of the coil with AC transport currents (20, 40, and 60 A) but an external DC magnetic field of 100 mT

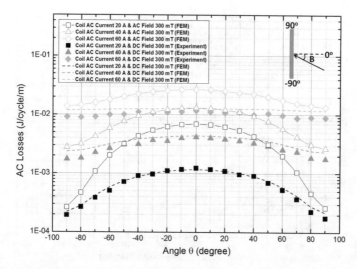

**Fig. 8.19** AC loss simulation of the coil with AC transport currents (20, 40, and 60 A) and the angular dependence of an external AC magnetic field of 300 mT, and its comparison to the AC loss experiment and simulation of the coil with AC transport currents (20, 40, and 60 A) but an external DC magnetic field of 300 mT

(20, 40, and 60 A) and the angular dependence of an external AC magnetic field of 100 mT, and its comparison to the AC loss experiment and simulation of the coil with an AC transport current (20, 40, and 60 A) but an external DC magnetic field of 100 mT. It can be noted for all the three AC transport current cases, that the AC loss difference between the external AC and DC magnetic fields at 100 mT was quite small when the field orientation angle, $\theta$, was $-90°$ or $90°$. For an AC transport current of 20 A, the difference became larger when angle $\theta$ was $0°$, but the difference was not significant for higher AC transport currents, e.g. the difference was only 15% for a 60 A transport current.

By contrast, Fig. 8.19 plots the AC loss simulation of the coil with AC transport currents (20, 40, and 60 A) and the angular dependence of an external AC magnetic field of 300 mT, and its comparison to the AC loss experiment and simulation of the coil with the same AC transport currents but an external DC magnetic field of 300 mT. A similar phenomenon can be seen, that when the field orientation angle $\theta$ was $-90$ or $90°$, the AC loss difference between the external AC and DC magnetic fields at 300 mT was slight. However, the difference significantly increased when the angle $\theta$ reached $0°$, for all the three transport current cases, e.g. 5.9 times greater for a 20 A transport current case, and 2.1 times greater for a 60 A transport current case. Therefore, one can summarise that for an HTS coil with an AC transport current and in the presence of an external DC or AC magnetic field with a given magnitude, an external AC magnetic field has a greater impact on its overall AC loss and angular dependence, particularly with stronger magnetic intensity fields.

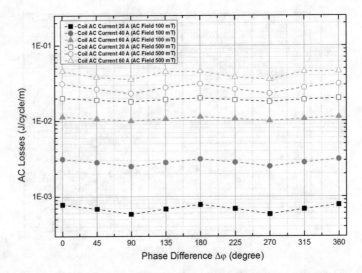

**Fig. 8.20** Phase difference $\Delta\varphi$ between the AC transport current and the external AC magnetic field (from 0 to 360°): AC loss simulation of the coil with an AC transport current (20, 40, and 60 A) and two sets of external AC magnetic field, 100 and 500 mT, perpendicular to the HTS tape surface of the coil

### 8.6.3  Phase Difference $\Delta\varphi$ Analysis with AC Transport Current and External AC Magnetic Field

The above analyses were based on the assumption that both the AC transport current and the external AC magnetic field were in phase. However, if there is a phase difference $\Delta\varphi$, namely, the AC transport current is leading or lagging behind the external AC magnetic field, the AC loss situation changes. Figure 8.20 illustrates the AC loss simulation of the coil with an AC transport current (20, 40, and 60 A) and two sets of external AC magnetic fields of 100 and 500 mT, perpendicular to the HTS tape surface of the coil. The phase difference between the AC transport currents and the external AC magnetic field, $\Delta\varphi$, was increased from 0 to 360°. As shown in Fig. 8.20, the phase shift between AC field and AC transport current has some influence on the AC losses of the coil, which depends on the amplitude of current and field.

From Fig. 8.20, it can be seen that for all the cases, the cycle of AC loss changes with $\Delta\varphi$ was 180°, and intuitively all the maximums occurred at values of $\Delta\varphi$ equal to 0, 180, and 360°, while all the minimums intuitively occurred at values of $\Delta\varphi$ equal to 90 and 270°. To be more precise, Fig. 8.21 depicts the AC loss simulation of the coil with an AC transport current (20, 40, and 60 A) and only one stronger external AC magnetic field of 500 mT, perpendicular to the HTS tape surface of coil, but with the phase difference $\Delta\varphi$ from 0 to 180°. It can be observed that, for the case of the 20 A transport current, the maximum loss happened when $\Delta\varphi$ was 0 and 180, and the minimum loss happened when $\Delta\varphi$ was 90. However, this kind of

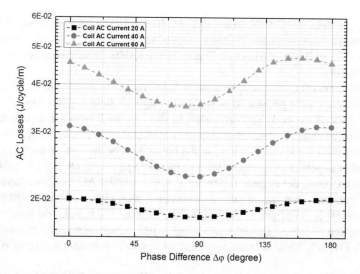

**Fig. 8.21** Phase difference $\Delta\varphi$ between the AC transport currents and the external AC magnetic field (from 0 to 180°): AC loss simulation of the coil with an AC transport current (20, 40, and 60 A) and an external AC magnetic field of 500 mT perpendicular to the HTS tape surface of the coil

"90° symmetry" of loss characteristics changed when the AC transport increased to 40 and 60 A. As presented in Fig. 8.21, the peak loss of 0.03118 (J/cycle/m) of the 40 A transport current case occurred when $\Delta\varphi$ was approximately 170°, and the peak loss of 0.04717 (J/cycle/m) of the 60 A transport current case occurred when $\Delta\varphi$ was approximately 160°. This kind of peak-shift phenomenon was more obvious with relatively stronger external AC magnetic fields, which is consistent with the measurements presented in [14].

## 8.7   Summary

A comprehensive study on AC losses in a circular HTS double pancake coil was carried out from both experiment and simulation. The experiment was established to measure the AC loss from a $2 \times 18$ circular double pancake coil using the electrical method. In order to enhance the consistency with the real circular coil used in the experiment, a 2D axisymmetric $H$-formulation model was built using the FEM package in COMSOL Multiphysics.

Three scenarios were analysed which can cause AC loss in an HTS coil: (1) AC transport current & DC magnetic field, (2) DC transport current & AC magnetic field, and (3) AC transport current & AC magnetic field. Moreover, the different orientation angles $\theta$ of the HTS coil under the magnetic field were studied for each scenario. For scenario (3), the impact of the relative phase difference between the

AC current and the AC field on the total AC loss of the coil was analysed. In short, a systematic current/field/angle/phase dependent AC loss ($I$, $B$, $\theta$, $\Delta\varphi$) study of a circular HTS coil was carried out. These are the results summarised by experiment and simulation.

For Scenario (1), AC transport current & DC magnetic field, both measurements and simulation were performed to study the magnetic field angular dependency on the AC loss of the coil. The difference in AC loss with each set of external DC fields became smaller when the transport current increased to a high value, which is consistent with the results reported in the literature. For the same transport current, the angular dependence increased with an increasing external DC field. For the same external DC field, the angular dependence with a smaller AC transport current was more significant than the angular dependence with a higher AC transport current. These results could be due to the fact that the external DC field is less influential on the reduction of critical current when a higher AC transport current is conducting in the coil. Overall, the simulation results showed good agreement with experimental results.

For Scenario (2), DC transport current & AC magnetic field, a simulation was performed for the AC loss in the coil related to $I/B/\theta$. Two phenomena were identified, similar to Scenario (1): (i) for the same transport current, the angular dependence increased with an increasing external field; (ii) for the same external field, the angular dependence with a smaller transport current was more significant than with a higher transport current. However, in the presence of the same magnetic field, increasing the DC transport current to a high value did not change the angular dependence too much, which is different from scenario (1). The AC loss increment tendency of scenario (2) was different from scenario (1), due to the fact that the AC losses were essentially caused by two different sources: AC transport current and AC external magnetic field.

For Scenario (3), AC transport current & AC magnetic field, a simulation was carried out to study the AC loss in the coil related to $I/B/\theta$. Again, it was found that the two phenomena, (i) and (ii), mentioned above occurred in all three scenarios. The angular dependency was compared with the experimental and simulation results from Scenario (1): for the HTS coil conducting an AC transport current in the presence of the same magnitude of external DC or AC magnetic field. The case of an external AC magnetic field presented a greater impact on the overall AC loss and angular dependence, especially with a stronger AC field. The effect of the phase difference $\Delta\varphi$ between AC transport currents and external AC magnetic field was investigated. The phase difference $\Delta\varphi$ had a greater impact on the total AC losses either with a larger AC transport current, or with a stronger external AC field. For further phase difference $\Delta\varphi$ analysis, the "90° symmetry" characteristics of AC loss changed when a larger AC transport current or a stronger external AC field were applied. These findings are consistent with reports in the literature.

Both the experiment and simulation were carried out in Scenario (1), and good consistency of experiment and simulation was presented in terms of AC loss magnitude, tendency, and angular dependence. The simulation results of Scenario (2) and

Scenario (3) were reasonable and consistent with previous work presented in the literature. Therefore, we believe that the coil model is able to produce convincing results throughout this multihinscenario study. A powerful coil model has the potential to efficiently compute the AC loss from various complex conditions. To summarise, a systematic study on AC loss from a circular HTS coil has been presented, and the methods and results of this study could be helpful for future analysis and design on HTS AC systems.

# References

1. D. Larbalestier, A. Gurevich, D.M. Feldmann, A. Polyanskii, High-Tc superconducting materials for electric power applications. Nature **414**, 368–377 (2001)
2. T. Tosaka, K. Koyanagi, K. Ohsemochi, M. Takahashi, Y. Ishii, M. Ono, H. Ogata, K. Nakamoto, H. Takigami, S. Nomura, Excitation tests of prototype HTS coil with Bi2212 cables for development of high energy density SMES. IEEE Trans. Appl. Supercond. **17**(2), 2010–2013 (2007)
3. B. Li, C. Li, F. Guo, Application studies on the active SISFCL in electric transmission system and its impact on line distance protection. IEEE Trans. Appl. Supercond. **25**(2) (2015)
4. B. Li, C. Li, F. Guo, Y. Xin, Overcurrent protection coordination in a power distribution network with the active superconductive fault current limiter. IEEE Trans. Appl. Supercond. **24**(5) (2014)
5. B. Shen, J. Li, J. Geng, L. Fu, X. Zhang, H. Zhang, C. Li, F. Grilli, T.A. Coombs, Investigation of AC losses in horizontally parallel HTS tapes. Supercond. Sci. Technol. **30**(7), Art. no. 075006 (2017)
6. W.V. Hassenzahl, D.W. Hazelton, B.K. Johnson, P. Komarek, M. Noe, C.T. Reis, Electric power applications of superconductivity. Proc. IEEE **92**(10), 1655–1674 (2004)
7. B. Shen, L. Fu, J. Geng, H. Zhang, X. Zhang, Z. Zhong, Z. Huang, T. Coombs, Design of a superconducting magnet for lorentz force electrical impedance tomography. IEEE Trans. Appl. Supercond. **26**(3), Art. no. 4400205 (2016)
8. B. Shen, L. Fu, J. Geng, X. Zhang, H. Zhang, Q. Dong, C. Li, J. Li, T. Coombs, Design and simulation of superconducting Lorentz force electrical impedance tomography (LFEIT). Phys. C Supercond. **524**, 5–12 (2016)
9. F. Grilli, E. Pardo, A. Stenvall, D.N. Nguyen, W. Yuan, F. Gömöry, Computation of losses in HTS under the action of varying magnetic fields and currents. IEEE Trans. Appl. Supercond. **24**(1), 78–110 (2014)
10. B. J. Parkinson, R. Slade, M. J. Mallett, V. Chamritski, Development of a cryogen free 1.5 T YBCO HTS magnet for MRI. IEEE Trans. Appl. Supercond. **23**(3) (2013)
11. N. Amemiya, K. Miyamoto, S.-I. Murasawa, H. Mukai, K. Ohmatsu, Finite element analysis of AC loss in non-twisted Bi-2223 tape carrying AC transport current and/or exposed to DC or AC external magnetic field. Phys. C Supercond. **310**(1), 30–35 (1998)
12. M. Ciszek, O. Tsukamoto, N. Amemiya, M. Ueyama, K. Hayashi, Angular dependence of AC transport losses in multifilamentary Bi-2223/Ag tape on external DC magnetic fields. IEEE Trans. Appl. Supercond. **9**(2), 817–820 (1999)
13. T. Chiba, Q. Li, S. Ashworth, M. Suenaga, P. Haldar, Angular dependence of ac losses at power frequencies for a stack of Bi2223/Ag tapes. IEEE Trans. Appl. Supercond. **9**(2), 2143–2146 (1999)
14. D. Nguyen, P. Sastry, G. Zhang, D. Knoll, J. Schwartz, AC loss measurement with a phase difference between current and applied magnetic field. IEEE Trans. Appl. Supercond. **15**(2), 2831–2834 (2005)

15. F. Grilli, F. Sirois, V.M. Zermeno, M. Vojenčiak, Self-consistent modeling of the Ic of HTS devices: How accurate do models really need to be? IEEE Trans. Appl. Supercond. **24**(6), Art. no. 8000508 (2014)
16. F. Grilli, V.M. Zermeno, E. Pardo, M. Vojenčiak, J. Brand, A. Kario, W. Goldacker, Self-field effects and AC losses in pancake coils assembled from coated conductor Roebel cables. IEEE Trans. Appl. Supercond. **24**(3) (2014)
17. F. Grilli, S.P. Ashworth, Measuring transport AC losses in YBCO-coated conductor coils. Supercond. Sci. Technol. **20**(8), 794–799 (2007)

# Chapter 9
# Investigation of AC Losses in Horizontally Parallel HTS Tapes

## 9.1 Introduction

This chapter presents an AC loss study of horizontally parallel HTS tapes. Racetrack and round HTS coils are the most common topologies used in many superconducting applications. In these coils, layers of HTS tape are closely packed, and carry current in the same direction. When observed from a 2D cross-section of the HTS coil, the tapes are stacked on top of each other. This configuration gives rise to large AC losses [1], due to the concentration of the perpendicular component of the magnetic field impinging on the HTS tapes [2]. This is the main reason why the AC losses in each turn of the HTS coil are larger than the AC losses in an equivalent single layer of HTS tape.

It is therefore interesting to use the geometry of horizontally parallel HTS tapes (shown in Fig. 9.1, sometimes also indicated as an x-array), which is a proper way to prevent high AC losses, due to the reduction of the perpendicular component of the magnetic field experienced by the tapes. The parallel HTS tape structure has been reported to potentially reduce the total AC losses in HTS power applications such as superconducting transformers [3].

In the literature there are several works dedicated to the AC losses of interacting parallel tapes. Müller [4] and Brambilla et al. [5] proposed analytical and numerical solutions for an x-array composed of an infinite number of tapes, respectively. Nakamura et al. performed an AC loss study of YBCO adjacent tapes [6]. Jiang et al. measured the transport AC losses in single and double layer parallel HTS tape arrays [7]. Lee et al. investigated the AC loss of parallel tapes with unbalanced current distribution [3]. Ogawa et al. measured and simulated the AC losses from an assembled conductor whose configuration was built as parallel HTS tapes [8]. There are other works regarding AC losses of two interacting parallel tapes [9, 10]. However, in the literature there is no detailed comparison of experiments and

B. Shen, *Study of Second Generation High Temperature Superconductors: Electromagnetic Characteristics and AC Loss Analysis*, Springer Theses, https://doi.org/10.1007/978-3-030-58058-2_9

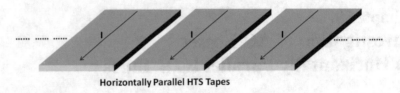

**Horizontally Parallel HTS Tapes**

**Fig. 9.1**  Configuration of multiple horizontally parallel HTS tapes

simulations for the AC losses of horizontally parallel tapes. A calculation of the
AC losses for the case of a large but finite number of tapes is also missing. This
chapter proposes the concept of using horizontally parallel HTS tapes, as schematically illustrated in Fig. 9.1. Firstly, three horizontally parallel tapes were used as the
basic configuration, and each parallel tape carried the current with the same amplitude and direction. Secondly, the experiment of this three–tape configuration was
set up and the AC losses from both the middle and end tapes were measured with
increasing gap distance. These measurements were compared with $H$-formulation
COMSOL simulations. A new parameter was proposed, $N_s$, a turning point for the
number of tapes, to divide Stage 1 and Stage 2 for the AC loss study of horizontally
parallel tapes. For Stage 1, $N < N_s$, the total average losses per tape increased with
the increasing number of tapes. For Stage 2, $N > N_s$, the total average losses per tape
began to decrease with the increasing number of tapes. By using simulation from
COMSOL, the cases of increasing parallel tapes were modelled, and then proposed
an empirical relation for the total average AC losses per tape in Stage 1. Larger
numbers of parallel tapes were also simulated to verify the approximate value of $N_s$.
Finally, the physical reason for why $N_s$ exists was analysed, and the results were
compared to the study from Jiang et al. on x-array HTS tapes [7].

## 9.2  Experiment Set-Up

Figure 9.2a presents the schematic of the electrical method used to measure the AC
losses of parallel HTS tapes, which is similar to the measurement schematic for the
single tape in Chap. 7.

The HTS tapes used in this experiment were SuperPower SF12100 (stabilizer-
free), 12 mm wide. The self-field critical current $I_c$ was measured to be approximately
300 A. As shown in Fig. 9.2a, on the secondary side, three horizontally parallel HTS
tapes (20 cm length) were fixed on a plastic board, with the aid of low temperature
KAPTON tapes. The tapes were connected in series. The copper cables for the series
connection were far away from the tapes (in order to reduce their field effect on the
tapes) and are not shown in Fig. 9.2a.

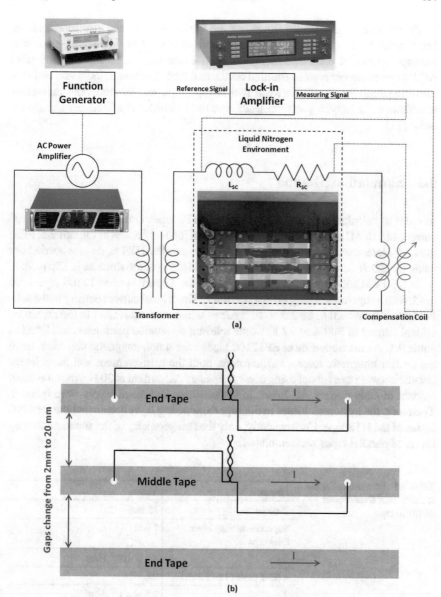

**Fig. 9.2** **a** Experimental schematic of AC loss measurements using the electrical method; **b** Graphical instruction of AC loss measurements on parallel HTS tapes

Figure 9.2b schematically illustrates the wiring for AC loss measurements on the parallel HTS tapes. First, AC losses were measured in the middle tape, with the gap distance of each tape increased from 2 mm to 20 mm, and then the same AC loss measurements were fulfilled on the end tape. For both middle and end tape measurements, the length between two voltage taps was 90 mm. The measuring technique of "8" glyph potential leads was used in order to effectively reduce the noise [11].

## 9.3   Simulation Method

In order to calculate the losses from parallel HTS tapes under the action of an AC current and an AC magnetic field, the $H$-formulation was chosen as the suitable FEM method. Here, the AC loss calculation from the parallel HTS tapes was carried out using the 2D $H$-formulation by COMSOL Multiphysics, the same as in Chap. 7.

In the FEM model, the real dimensions of the SuperPower SF12100 tape were used, with a superconducting layer 1 $\mu$m thick. The critical current density in the self-field was determined to be $2.5 \times 10^{10}$ A/m$^2$, which was equivalent to the measured critical current of 300 A at 77 K. Some relevant simulation parameters are listed in Table 9.1. As the SuperPower SF12100 tapes have a non–magnetic substrate, there are no ferromagnetic losses. Furthermore, both the measurement and the relevant simulation were carried out using a low frequency AC current at 20 Hz, thus the small amount of eddy-current AC losses in the substrate and metal layers were ignored. Therefore, the hysteresis losses in the superconducting layer dominated the total AC losses of the HTS tape. Consequently, only the real geometry of the superconducting layers of parallel tapes was simulated.

**Table 9.1** Parameters for the simulation of SuperPower SF12100 tape

| Parameters | Value |
| --- | --- |
| Tape width | 12 mm |
| Superconducting layer thickness | 1 $\mu$m |
| $\mu_0$ | $4\pi \times 10^{-7}$ H/m |
| $n$ (E–J power law factor) | 25 |
| $J_{c0}$ | $2.5 \times 10^{10}$ A/m$^2$ |
| $E_0$ | $10^{-4}$ V/m |
| Gap of tapes | 2, 4, 6, 8, 10, 12, 14, 16, 18, 20 mm |

## 9.4  Results and Discussion from Three Horizontally Parallel Tape Cases

### 9.4.1  Influence of the Gap Distance on AC Losses

The AC transport current applied to the three parallel HTS tapes was increased from 60 to 150 A (peak values). Before the measurement was taken on the parallel tapes, the AC losses of an individual single tape were also measured and used for reference.

As demonstrated in Fig. 9.3, when the gap distance increased from 2 to 20 mm, the AC losses of the middle tape gradually increased. For the 2 mm gap case, the losses in the middle tape were $1.99 \times 10^{-4}$ J/cycle/m at a current of 150 A peak, which is 3 times lower than the losses of the individual tape with the same transport current. When the gap was 4 mm, the AC losses in the middle tape were half ($2.96 \times 10^{-4}$ J/cycle/m) of those of the individual tape. From the 4 mm gap case onward to the 20 mm gap case, the change in rate of AC losses in the middle tape became slower.

Figure 9.4 presents the AC loss measurements on the end tape of three parallel tapes. When the gap distance was 2 mm, the AC losses measured on the end tape were $2.1 \times 10^{-3}$ J/cycle/m with the transport current 150 A peak, which was more than 3.5 times higher than the losses of the individual tape. With the gap increasing from 2 to 20 mm, the AC losses on the end tape started to decrease and to approach the loss value of the individual tape, and the decreasing speed slowed down as well.

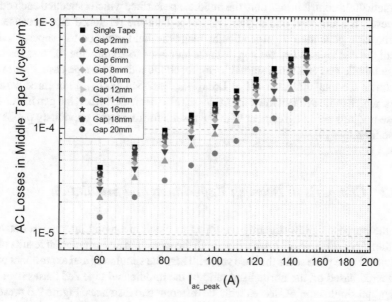

**Fig. 9.3**  Measured AC losses in the middle tape of the three parallel tapes (Gap changed from 2 to 20 mm)

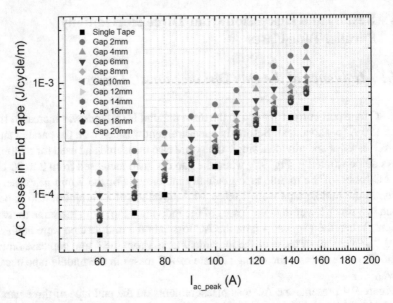

**Fig. 9.4** Measured AC losses in the end tape of the three parallel tapes (Gap changed from 2 to 20 mm)

As it will be shown later, the reason why the middle tape had lower AC losses than the reference individual one was that the two end tapes created perpendicular magnetic field contributions onto the middle tape surface which cancelled each other. However, the end tape had higher AC losses than the reference tape because the middle tape generated a perpendicular magnetic field contribution superposed to that of the end tape. When the gap distance increased to 20 mm, the AC loss features of the middle and end tapes started to approach the individual tape case. Figure 9.5 illustrates the total AC losses ($1 \times \text{Loss}_{\text{mid tape}} + 2 \times \text{Loss}_{\text{end tape}}$) of three parallel tapes, which indicates that the total AC losses also decreased when the gap increased. These findings are in agreement with the results for two tapes previously published in the literature [9].

## 9.4.2 Comparison Between Experiment and Simulation

For the purpose of investigating the effect of electromagnetic interaction between parallel tapes on their AC losses, first, both the measured and simulation results of an individual tape were set as the base values. Then, the simulation and experiment were compared based on the normalised ratio of the middle/end tape AC losses over the individual single tape AC losses with the different gap distances. Figure 9.6 presents such a comparison for a transport current of 150 A at 20 Hz. In general, there is a good agreement between the experiment and the simulation. However, there were

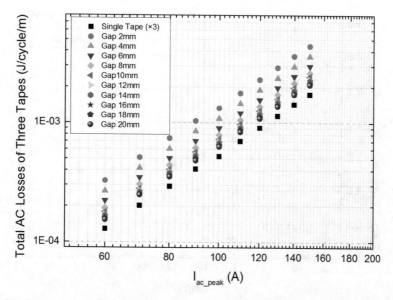

**Fig. 9.5**  Total measured AC losses from the three parallel tapes (Gap changed from 2 to 20 mm)

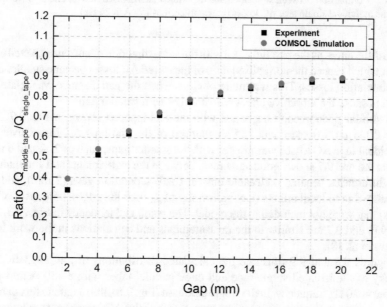

**Fig. 9.6**  Comparison between simulation and experiment: normalised ratio of middle tape AC losses over the individual single tape AC losses (transport current 150 A at 20 Hz)

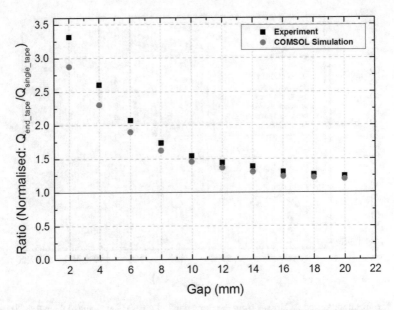

**Fig. 9.7** Comparison between simulation and experiment: normalised ratio of end tape AC losses over the individual single tape AC losses (transport current 150 A at 20 Hz)

some differences as the parallel tapes with the same direction of current attracted each other, which caused the normalised ratio of measured AC losses became smaller than the simulation losses. This was more obvious when the gap distance was relatively small because the attractive effect was stronger with smaller gaps.

A similar phenomenon happened with the end tape, as shown in Fig. 9.7. The normalised ratio (experiment and simulation) of the end tape AC losses over the individual tape AC losses was greater with the smaller gaps, such as 2, 4, and 6 mm. Figure 9.6 and 9.7 shows good agreement between the experiments and simulations. The differences became noticeable only at small separation gaps (2 mm), and are probably due to local effects (e.g. uniformity of $J_c$ near the edges, alignment of the tapes) that were not included in the model. The result and tendency in Figs. 9.3, 9.4, 9.5, 9.6, and 9.7 are similar to the measurements and calculations in the work from Ogawa et al. [8].

Figures 9.8 and 9.9 illustrate the calculated distributions of the magnetic flux density of the isolated single tape and three parallel tapes with a 150 A transport current at 20 Hz frequency. In the single tape case (Fig. 9.8), the magnetic flux density was greater near the two edges of the tape, and thus the AC losses became larger on the two sides of the tape. Compared to the case of three parallel tapes in Fig. 9.9, the magnetic flux density around the two sides of the middle tape was weaker, as the two end tapes generated perpendicular magnetic fields that cancelled each other, thus effectively shielding the middle tape in between. This was the reason why the AC losses were smaller in the middle tape, but as the gap increased, the shielding

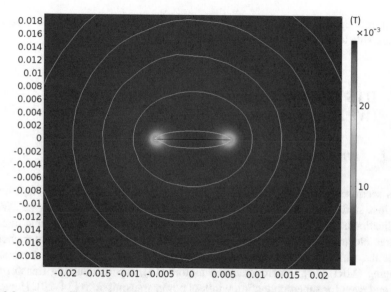

**Fig. 9.8** Magnetic flux density of an isolated single tape with a 150 A transport current at 20 Hz from the COMSOL simulation

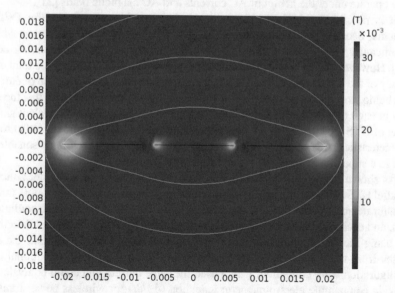

**Fig. 9.9** Magnetic flux density of three parallel tapes with a 150 A transport current at 20 Hz from the COMSOL simulation (gap distance 2 mm)

effect from the two end tapes became weaker, and the losses of the middle tape were comparable to losses from each end tape.

## 9.5 HTS Tapes for Possible Wireless Power Transfer (WPT) Applications

### 9.5.1 Introduction

This section presents the concept of using horizontally parallel HTS tapes with AC loss study, and the investigation into possible wireless power transfer (WPT) applications. High temperature superconductor (HTS) based coils are able to carry a large electrical current density. HTS coils are one of the most popular choices to be used in various superconducting applications such as Magnetic Resonance Imaging (MRI) [12], superconducting motors [13], superconducting transformers [3], and even for superconducting wireless power transmissions [14–20]. However, alternating current (AC) losses are the crucial problems for HTS applications when they operate under the action of AC currents and AC magnetic fields [21].

It is possible to apply superconductors into wireless power transfer (WPT) systems, because superconducting coils have theoretically zero resistance which significantly reduces the resistive losses from the transmitter and receiving coils [22]. However, the AC losses from the HTS coil still potentially affect the total efficiency of the WPT system. The HTS tapes are carrying the same direction of current and being closely packed in these coil structures. This is the main reason that the AC loss in each turn of the HTS coil is larger than the AC loss in an equivalent single layer of HTS tape, owing to the field dependency of the AC loss and the anisotropic characteristics of HTS tapes [2]. Therefore, for a WPT system, it is reasonable to design a superconducting part which has less AC losses.

As shown in Fig. 9.10, this section proposes the concept of using horizontally parallel HTS tapes, whose configuration could be potentially used for wireless power transfer devices. Three horizontally parallel tapes were used as the basic configuration, and each parallel tape carried a current with the same amplitude and direction. By using the $H$-formulation electromagnetic simulation from COMSOL, the AC losses from the parallel tapes configuration and those from the conventional coil configuration were compared, which used the same current and could equivalently provide comparable electromagnetic induction ($d\phi/dt$) for wireless power transfer in a certain cross-section.

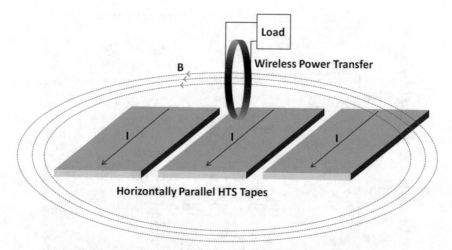

**Fig. 9.10** Configuration of multiple horizontally parallel HTS tapes, which could be potentially used for wireless power transfer applications

### 9.5.2 AC Losses in Three Parallel Tapes (Summary for WTP Analysis)

Figure 9.11 presents the measurements and simulation for the AC losses in a single tape, where Norris's analytical solutions for both the Strip and Ellipse cases are shown as the Pardo et al. [23]. The AC transport current applied to a single tape was increased from 60 to 150 A of peak value for the measurement, while the applied transport current was set from 60 to 300 A of peak value for the simulation. The frequency of the transport current was set to be 20 Hz. The AC losses from both the simulation and measurements were between the range of Norris Strip and Norris Ellipse, and the simulation results agreed with the experimental results.

For the three parallel HTS tapes, the AC losses in the middle, and then end tapes, were measured and simulated, with the gap distance between tapes increasing from 2 to 20 mm. In order to investigate the interactive effect of parallel tapes in a straightforward way, both the experimental and simulation results of a single tape were set as the base values. Similar to the analysis in Sect. 9.4, the simulation and measurements were converted to the normalised ratio: middle tape, end tape and total average AC losses (($1 \times$ Losses$_{mid tape} + 2 \times$ Losses$_{end tape}$)/3) over the isolated single tape AC losses. Figure 9.12 demonstrates the comparison of simulation and measurements based on normalised ratio of the middle tape, end tape, and total average over the single tape AC losses. The simulation results were generally consistent with the measured results.

From Fig. 9.12, it can be seen that the normalised ratio of the middle and end tapes approached "1" when the gap increased, which means that the interactive effect on the AC losses became less significant with larger gaps. This is similar to the previous

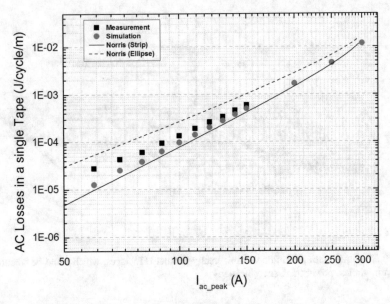

**Fig. 9.11** Comparison between measurement and simulation: AC losses in a single (isolated) tape, with reference to Norris's analytical solutions (Strip and Ellipse), at a transport current frequency of 20 Hz

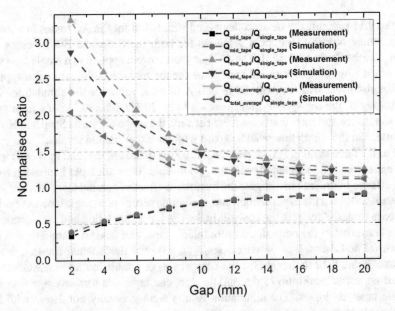

**Fig. 9.12** Comparison between simulation and measurement: normalised ratio of mid tape AC losses over a single (isolated) tape AC losses; normalised ratio of end tape AC losses over the single (isolated) tape AC losses; normalised ratio of total average AC losses over the single (isolated) tape AC losses; with a transport current of 150 A at 20 Hz

work [24]. As the AC losses in the end tapes dominated the total losses, the tendency of the total average AC losses was similar to the tendency of the end tape AC losses.

As shown in Fig. 9.12, the largest AC loss of the three parallel HTS tapes occurred with the 2 mm gap, whose total average AC loss was approximately twice that of the individual single tape from the simulation, and even slightly higher than the measurement. It could be calculated from the simulation that the total AC loss of the three parallel tapes was approximately 0.00311 J/cycle/m, with a gap distance of 2 mm, and a transport current of 150 A at 20 Hz.

### 9.5.3 Electromagnetic Induction (EMI) from Horizontally Parallel HTS Tapes

(a) *Electromagnetic induction simulation*

The horizontally parallel HTS tapes with proper electrical current were able to produce useful magnetic field patterns. In this section, the electromagnetic induction (EMI) around three parallel tapes was analysed using the *H*-formulation with the FEM package of COMSOL Multiphysics. As shown in Fig. 9.13, the electromag-

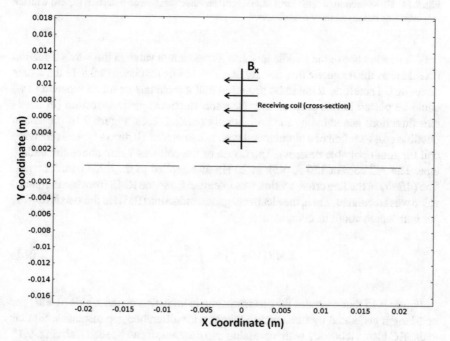

**Fig. 9.13** Cross-section of three parallel HTS tapes and electromagnetic induction ($d\phi/dt$) with the magnetic flux density in the x-axis direction

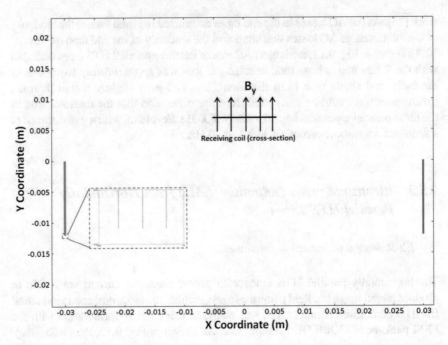

**Fig. 9.14** Cross-section of three turns of HTS coil and electromagnetic induction (d$\phi$/d$t$) with the magnetic flux density in the y-axis direction

netic induction above the middle tape has its maximum value in the x-axis direction (the same as the magnetic flux density, $B_x$, also has its maximum value in the x-axis direction). Therefore, it could be proposed that a rectangle or round receiving coil could be placed in this position to collect the electromagnetic induction (in the x-axis direction) generated by the horizontally parallel tapes. Figure 9.14 presents a possible cross-section of a circular coil with a diameter of 10 mm (a typical receiving coil for small portable devices). The centre of the coil was 7 mm above the middle tape. The AC current 150 A peak at 20 Hz was applied to the three parallel tapes. The (d$B$/d$t$) in the line cross-section was integrated, and the RMS (root mean square) value was calculated. Then, the electromagnetic induction (EMI) in the x-axis (d$\phi$/d$t$) per unit-length could be computed:

$$EMI\left(Vm^{-1}\right) = \int_l \frac{dB_x}{dt} \, dl \qquad (9.1)$$

Figure 9.15 demonstrates the electromagnetic induction on the x-axis (d$\phi$/d$t$) per unit-length generated by three parallel tapes with difference gap distances. Similar to the AC losses tendency with increasing gap distance, it can be seen from Fig. 9.15 that the electromagnetic induction per unit-length decreased when the gap increased,

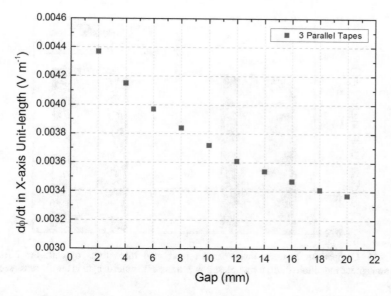

**Fig. 9.15** Electromagnetic induction ($d\phi/dt$) with the magnetic flux density in the x-axis direction from three parallel HTS tapes, with different gap distances

but the decreasing speed of electromagnetic induction per unit-length became slower when the gap distance increased to a high value.

(b)   *Comparison between three parallel tapes and three turns of coil*

Coils are one of the most common topologies for building a magnetic field, and create electromagnetic induction which is used in wireless power transfer systems. However, if coils are fabricated from superconducting tapes, the AC losses need to be taken into consideration because of the anisotropic characteristics and field dependency of HTS tapes.

Figure 9.14 illustrates the cross-section of three turns of an HTS coil. The real geometry of the tape with a superconducting layer of 1 μm was used, and each tape cross-section had a real tape separation thickness of 0.1 mm. There is a zoomed–in part of Fig. 9.14 which clearly shows the cross-section of the three turns of coil. Typically, for wireless power transfer, the transmitting coil and receiving coil are placed vertically parallel, in order to obtain more electromagnetic induction. As shown in Fig. 9.14, the receiving coil was to collect the electromagnetic induction (in the y-axis direction) generated from the three turns of coil. The same size of the receiving coil cross-section (10 mm diameter) was used as the three parallel tape case mentioned above, and was placed in the same position 7 mm above the transmitting system (HTS coil level). The same AC current of 150 A at 20 Hz was applied to the HTS coil.

Figure 9.16 presents a comparison of the electromagnetic induction per unit-length generated from three parallel tapes (2 mm gap) and three turns of coil, using

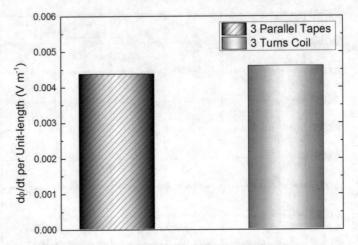

**Fig. 9.16** Comparison between three parallel tapes and three turns of coil: the effective electromagnetic induction (df/dt) for wireless power transfer per unit-length in the 2D cross-section

the same AC current and the same receiving coil cross-section with a diameter of 10 mm. For the three parallel tapes the electromagnetic induction per unit-length was 0.0435 Vm$^{-1}$, and for the three turns of coil the electromagnetic induction per unit-length was 0.0460 Vm$^{-1}$, which indicates that their difference (5%) is not significant. Figure 9.17 demonstrates the comparison of the total AC losses per unit-length from three parallel tapes (2 mm gap) and three turns of coil, which are 0.00311 J/cycle/m for the three parallel tapes, and 0.0151 J/cycle/m for three turns of coil. From Fig. 9.17

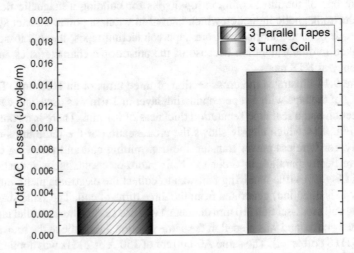

**Fig. 9.17** Comparison between three parallel tapes and three turns of coil: the total AC losses per unit-length in the 2D cross-section

it can be clearly seen that the difference in their total AC losses was approximately 5 times.

To summarise, for a particular case with a receiving coil with a cross-section 10 mm diameter above the transmitting system, the three horizontally parallel tapes can achieve a comparable electromagnetic induction per unit-length, but generated only 1/5 of the total AC losses compared to three turns of coil. However, the three parallel tapes are suitable when the receiving coil diameter is less than a certain value (15 mm), and the receiving coil is just above the transmitting system, whose configuration could be used for small scale portable devices charging on wireless power transfer base. For larger receiving coils, parallel tapes are no longer as efficient as the coil structure. This is because the electromagnetic induction decays sharply when the centre of the receiving coil is far away from the level of transmitting system. Moreover, the conventional vertically parallel coils may have better efficiency and quality of wireless power transfer. Nevertheless, the low total AC losses characteristic of multiple parallel tapes could be beneficial for the further investigation of electromagnetic applications using parallel HTS tapes configuration.

## 9.6   Total Average AC Losses from Increasing the Number of Horizontally Parallel Tapes

In order to analyse the efficiency of a certain system, the total AC losses should be considered. For the case of three parallel tapes as described in the previous sections, Fig. 9.18 presents the comparison of simulation and experimental results based on the normalised ratio of total average $((1 \times \text{Losses}_{\text{mid tape}} + 2 \times \text{Losses}_{\text{end tape}})/3)$ AC losses per tape over the individual single tape AC losses. As discussed in Sect. 9.5.2, the tendency was similar to the end tape case because the end tape losses dominated the total losses, especially when the gap distances were small. The normalised ratios approached unity when the gap increased to higher values, because the interactive effect of the parallel tapes on AC losses became less significant. These results are also similar to the work from Jiang et al. [7], which was carried out with narrower 4 mm tapes.

With the number of parallel tapes increasing, the total loss could consequently increase. However, knowing whether the total average AC loss per tape (total loss/n) would increase or decrease with an increasing number of tapes requires a dedicated investigation. Jiang et al. mentioned in [7] that two regimes can be distinguished: in regime 1 $(N < N_0)$, for a small number of tapes in the x-array, the total average AC losses per tape is higher than the AC losses of a single tape, due to the contribution of the tapes situated near the ends of the array; in regime 2 $(N > N_0)$, for a large number of tapes in the x-array, the total average AC losses per tape is smaller than the AC losses from a single tape. The relative weight of the AC losses of the end tapes decreases and the situation resembles gradually that of an infinite x-array, which has been analytically solved by Müller [4].

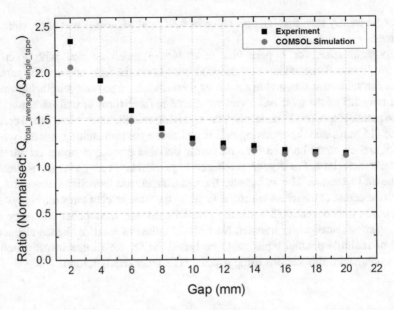

**Fig. 9.18** Comparison between simulation and experiment: normalised ratio of total average AC losses over the individual single tape AC losses with a transport current of 150 A at 20 Hz

The aforementioned viewpoint of Jiang et al. is reasonable, and both experimental and simulation results show that, for a small number of tapes, the total average AC losses per tape are greater than the AC losses of a single tape. In this study, the transition between Regime 1 and Regime 2 is not a sudden change at a number $N_0$. A new parameter was proposed, $N_s$, a certain number of parallel tapes, to divide Stage 1 and Stage 2. For Stage 1 ($N < N_s$), the total average losses per tape increased with the increasing number of tapes. For Stage 2 ($N > N_s$), the total average losses per tape started to decrease with the increasing number of tapes.

### 9.6.1  Stage 1($N < N_s$)-Total Average Losses per Tape

For Stage 1, the experiment and simulation for the three parallel tapes were performed. Then, cases with more tapes (7, 11, and 15) were simulated. Figure 9.19 presents the variation of the losses tape-by-tape for the case of seven tapes with a gap distance of 2 mm and a transport current of 150 A at 20 Hz. It can be seen that the losses of the middle tapes were relatively small while the losses of the end tapes still dominated the total AC losses with this small gap (2 mm). As shown in Fig. 9.20, the simulations were executed for the cases of 3, 7, 11, and 15 tapes, and the normalised ratio of the total average AC losses per tape over the individual single tape AC losses were calculated for different gap distances. From Fig. 9.20, it can be noted that the normalised ratio decreases to approach unity with the increasing gap

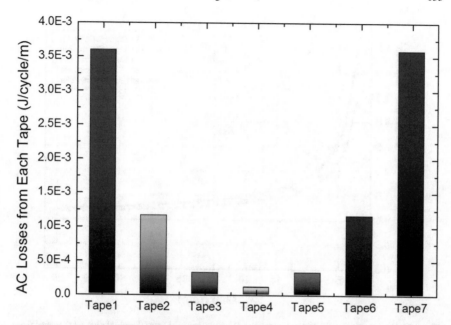

**Fig. 9.19** Variation of the losses tape-by-tape for the case of seven tapes with a gap distance of 2 mm and a transport current of 150 A at 20 Hz

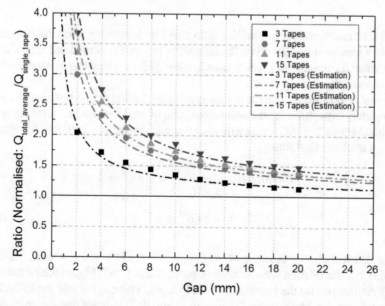

**Fig. 9.20** Normalized ratio of total average AC losses per tape over the individual single tape AC losses from the COMSOL simulation, compared to the numerical estimation calculated using Eqs. (9.2) and (9.3) (cases of 3, 7, 11, and 15 parallel tape)

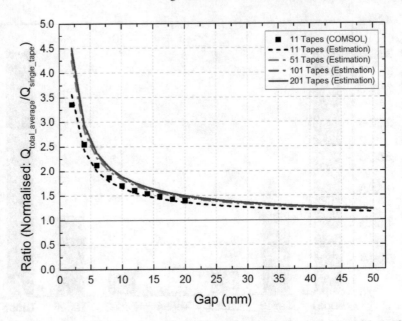

**Fig. 9.21** Normalised ratio of total average AC losses per tape over the single tape AC losses from Stage 1 numerical estimation calculated using Eqs. (9.2) and (9.3) (cases of 11, 51, 101, and 201 parallel tapes), also compared with the COMSOL simulation of 11 parallel tapes case

for all four cases, and the normalised ratio also kept increasing when the number of parallel tapes went up. However, the increasing rate of the normalised ratio slowed down when the number of parallel tapes became higher, which can be seen from the four cases in Fig. 9.20.

In order to efficiently find the boundary of Stage 1 other than simulating all cases, an empirical expression for the normalised ratio of total average AC losses over the individual single tape AC losses with different gap distances is proposed for Stage 1 (using a mathematical fitting):

$$Ratio(Normalised) = a \cdot Gap^b + 1 \tag{9.2}$$

$$a = c \cdot N^d + e \tag{9.3}$$

where constants $b = -0.852$, $c = -9.036$, $d = -0.6212$, and $e = 6.66$. Variable "a" also has the relation with number of parallel tapes "N". Figure 9.20 shows the fitting estimations for the cases of 3, 7, 11, 15 tapes, which agree with the COMSOL simulation results. Mathematically, Eqs. (9.2) and (9.3) indicate that the normalised ratio converges to unity not only with increasing gap distance but also with the increasing number of parallel tapes in Stage 1.

Figure 9.21 presents the numerical estimation calculated using Eqs. (9.2) and (9.3): the normalised ratio of the total average AC losses over the individual single tape AC losses with gap distances up to 50 mm. The number of parallel tapes was increased from 11, to 51, and from 101 to 201. As shown in Fig. 9.21, the calculation of 11 tapes agreed with the COMSOL simulation result of 11 tapes. It can be seen from Fig. 9.21 that the difference between 11 and 51 tapes cases was quite substantial, while the differences between 51, 101 and 201 tapes cases were fairly small. It could be deduced that the total average AC losses per tape would start to saturate when the number of tapes reaches certain level in Stage 1.

### 9.6.2 Stage 2(N > Ns)-Total Average Losses per Tape

Based on the numerical relation and simulation results from Stage 1, the increasing trend of the normalised ratio for the total average AC losses per tape would saturate for a number of tapes in the order of a few hundreds. Then, the number of parallel tapes increased with bigger step in order to discover the value of $N_s$. As shown in Fig. 9.22, the cases of 7, 15, 27, 51, 75, 101, 201, 251, 301, 351, 401, 501, 601, 801, and 1601 tapes were simulated with a constant gap distance of 2 mm, and a transport current of 150 A at 20 Hz. It can be seen that the turning point, $N_s$, was

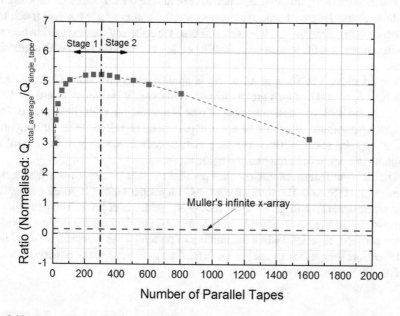

**Fig. 9.22** Cases of 7, 15, 27, 51, 75, 101, 201, 251, 301, 351, 401, 501, 601, 801, and 1601 parallel tapes with a constant gap distance of 2 mm and a transport current of 150 A at 20 Hz using COMSOL simulation, and the analysis of $N_s$, Stage 1 and Stage 2, with reference of Müller's analytical result for infinite tapes

around the case of 301 tapes (which corresponds to a total width of the array of about 4.2 m). From this point onward, the total average AC losses per tape started to decrease with the further increasing number of parallel tapes. The maximum number of tapes simulated is 1601, and for this case the total average AC losses per tape has decreased to a value similar to that of seven tapes. The normalised ratio of the 1601 tape case was still a long way, not only from the Müller's infinite x-array with same gap distance (shown with a dashed line in Fig. 9.22), but also, from the losses of the individual tape.

### 9.6.3  Summary

Based on the results shown above, the total average losses per tape would firstly increase, saturate, and then decrease with the incremental number of horizontally parallel tapes. In Stage 1, the parallel tapes interacted more strongly with each other when the number of tapes was small. Although the middle tapes had lower losses than a single tape, the end tapes dominated the total losses because of the strong magnetic fields around them, and the total average losses per tape increased with increasing tapes. There was a turning point, $N_s$, dividing the Stages 1 and 2, which could distinguish when the total average losses per tape would increase or decrease with increasing tapes. In Stage 2, when the amount of parallel tapes became higher, the total planar length of tape system was longer, and the far tapes had much less influence. Therefore, the AC losses contributed from the end tapes started to become less important, and the total average losses per tape began to decrease with increasing tapes.

The conclusion of Jiang et al. on $N_0$ is applicable to this study. Comparing $N_s$ with $N_0$, $N_0$ should be much greater than $N_s$. However, it is difficult to use the FEM simulation to achieve massive amount of tape elements, due to the limitation of PC's computing capability. $N_s$ could be a useful parameter to determine the turning point where the total average losses per tape start to reduce, which will be helpful for designing a system of horizontally parallel tapes.

The AC loss feature of horizontally parallel tapes is different from the stacked tapes in an HTS coil. The total average AC losses per tape still go up observably even when the HTS coil geometry increases to hundreds of turns. The geometry of horizontally parallel tapes could be able to maintain or even reduce the total average losses per tape, especially when the number of parallel tapes increases to high level, and therefore this geometry could be potentially applied to superconducting devices such as HTS transformers.

## 9.7  Summary

A study on AC losses of horizontally parallel HTS tapes has been presented. AC loss measurements of the middle and end tapes of three parallel tapes were carried out using the electrical method. The AC losses from an individual single tape were also measured and used as a reference. The middle tape had less AC losses than the reference individual tape, due to the reduction of the perpendicular field component at its edges. On the contrary, the end tape had larger AC losses than the reference individual tape due to the substantial perpendicular field component at one of its edges. The interactive effect on the AC losses from the middle and end tapes became less significant when the gap distance increased, and the loss value started to approach that of the individual tape. The simulation of the parallel-tape experiment was executed by the 2D $H$-formulation on the FEM platform of COMSOL Multiphysics. The simulation results fit the experimental results on the AC loss features of middle and end tapes with increasing gap distance.

The concept of using horizontally parallel HTS tapes for possible wireless power transfer (WPT) applications has been presented, together with their AC loss analysis. A receiving coil with 10 mm diameter cross-section was able to collect a comparable electromagnetic induction (per unit-length) generated by either three parallel tapes or by three turns HTS of coil, with the same AC transport current and the same position of receiving coil 7 mm above the transmitting system. The total AC losses from the three parallel tapes case was only 1/5 of the total AC losses from the three turns coil case. However, the electromagnetic induction decays greatly if the centre of the receiving coil moves far away from the level of the transmitting system, and the conventional vertically parallel coils present better wireless power transfer efficiency in a more spacious region. However, the configuration of multiple horizontally parallel tapes system has the obvious advantage of lower total AC losses than multiple tapes closely stacked in a conventional coil. The horizontally parallel tapes could create useful patterns of electromagnetic induction, and could potentially be used in wireless power transfer system for portable devices.

The total average losses per tape of horizontally parallel tapes have been investigated. A new parameter was proposed, $N_s$, which is the number of parallel tapes that divides Stages 1 and 2. In Stage 1, $N < N_s$, the parallel tapes contributed more interaction to each other, and experience high peak magnetic fields around the end tapes, which caused large AC losses in the end tapes, dominating the total losses. As a consequence, the total average losses per tape increased with increasing number of tapes. In Stage 2, $N > N_s$, the more distant tapes had much less impact on the total losses as the total planar length of the tape system became longer, and the AC losses contributed from the end tapes tended to become relatively negligible. Therefore, the total average losses per tape started to decrease with an increasing number of tapes. In order to efficiently find the boundary of Stage 1 other than using COMSOL to simulate all cases, a mathematical relation between the normalised ratio of the total average AC losses per tape and different gap distances has been proposed. According to this expression, the normalised loss ratio converges to one (i.e. the average loss per tape converges to that of an isolated tape) with increasing gap distance, but also with an increasing number of parallel tapes. According the study of Jiang et al., it could be estimated that the total average AC losses would approach the Müller's analytical results for infinite tapes with number of tapes $N_0$, but $N_0$ should be much larger than $N_s$. The turning point, $N_s$, could be a useful factor to resolve where the total average losses per tape start to reduce, which will be helpful for designing a system of horizontally parallel tapes. The geometry of horizontally parallel tapes could be potentially applied to superconducting devices like HTS transformers, because this structure is able to maintain and even reduce the total average AC losses per tape with large number of tapes.

# References

1. F. Grilli, S.P. Ashworth, Measuring transport AC losses in YBCO-coated conductor coils. Supercond. Sci. Technol. **20**(8), 794–799 (2007)
2. E. Pardo, F. Grilli, Numerical simulations of the angular dependence of magnetization AC losses: coated conductors, Roebel cables and double pancake coils. Supercond. Sci. Technol. **25**(1) (2011)
3. S. Lee, S. Byun, W. Kim, K. Choi, H. Lee, AC loss analysis of a HTS coil with parallel superconducting tapes of unbalanced current distribution. Phys. C, Supercond. **463**, 1271–1275 (2007)
4. K.-H. Müller, AC losses in stacks and arrays of YBCO/hastelloy and monofilamentary Bi-2223/Ag tapes. Physica C **312**(1), 149–167 (1999)
5. R. Brambilla, F. Grilli, D. N. Nguyen, L. Martini, F. Sirois, AC losses in thin superconductors: the integral equation method applied to stacks and windings. Supercond. Sci. Technol. **22**(7) (2009)
6. S. Nakamura, J. Ogawa, O. Tsukamoto, U. Balachandran, *AC transport current losses in YBCO tapes with adjacent tapes*, pp. 877–884
7. Z. Jiang, N. Long, M. Staines, Q. Li, R. Slade, N. Amemiya, A. Caplin, Transport AC loss measurements in single-and two-layer parallel coated conductor arrays with low turn numbers. IEEE Trans. Appl. Supercond. **22**(6) (2012)
8. J. Ogawa, S. Fukui, M. Yamaguchi, T. Sato, O. Tsukamoto, S. Nakamura, Comparison between experimental and numerical analysis of AC transport current loss measurement in YBCO tapes in an assembled conductor. Phys. C, Supercond. **445**, 1083–1087 (2006)
9. Z. S. Wu, Y. R. Xue, J. Fang, Y. J. Huo, L. Yin, *The influence of the YBCO tape arrangement and gap between the two tapes on AC losses*, pp. 205–206
10. K. Ryu, K. Park, G. Cha, Effect of the neighboring tape's AC currents on transport current loss of a Bi-2223 tape. IEEE Trans. Appl. Supercond. **11**(1), 2220–2223 (2001)
11. Y. Zhao, J. Fang, W. Zhang, J. Zhao, L. Sheng, Comparison between measured and numerically calculated AC losses in second-generation high temperature superconductor pancake coils. Phys. C, Supercond. **471**(21), 1003–1006 (2011)
12. J. R. Hull, Applications of high-temperature superconductors in power technology. Rep. Progr. Phys. **66**(11) (2003)
13. Z. Huang, H. S. Ruiz, Y. Zhai, J. Geng, B. Shen, T. Coombs, Study of the pulsed field magnetization strategy for the superconducting rotor. IEEE Trans. Appl. Supercond. **26**(4), Art. no. 5202105 (2016)
14. R.J. Sedwick, Long range inductive power transfer with superconducting oscillators. Ann. Phys. **325**(2), 287–299 (2010)

15. I.K. Yoo, et al., Design and analysis of high Q-factor LC resonant coil for wireless power transfer. Korean Inst. Inform. Technol. 9–116 (2014)
16. I.-S. Jeong, H.-S. Choi, M.-S. Kang, Application of the superconductor coil for the improvement of wireless power transmission using magnetic resonance. J. Supercond. Novel Magn. 28(2), 639–644 (2015)
17. I.-S. Jeong, Y.-K. Lee, H.-S. Choi, Characteristics analysis on a superconductor resonance coil WPT system according to cooling vessel materials in different distances. Phys. C, Supercond. 530, 123–132 (2016)
18. W. Zuo, S. Wang, Y. Liao, Y. Xu, Investigation of efficiency and load characteristics of superconducting wireless power transfer system. IEEE Trans. Appl. Supercond. 25(3) (2015)
19. A. Kurs, A. Karalis, R. Moffatt, J.D. Joannopoulos, P. Fisher, M. Soljačić, Wireless power transfer via strongly coupled magnetic resonances. Science 317(5834), 83–86 (2007)
20. B. Shen, J. Li, J. Geng, L. Fu, X. Zhang, H. Zhang, C. Li, F. Grilli, T. A. Coombs, Investigation of AC losses in horizontally parallel HTS tapes. Supercond. Sci. Technol. 30(7), Art. no. 075006 (2017)
21. G. Zhang, H. Yu, L. Jing, J. Li, Q. Liu, X. Feng, Wireless power transfer using high temperature superconducting pancake coils. IEEE Trans. Appl. Supercond. 24(3), Art. no. 4600505 (2014)
22. W. Norris, Calculation of hysteresis losses in hard superconductors carrying ac: isolated conductors and edges of thin sheets. J. Phys. D Appl. Phys. 3(4), 489–507 (1970)
23. E. Pardo, A. Sanchez, D.-X. Chen, C. Navau, Theoretical analysis of the transport critical-state ac loss in arrays of superconducting rectangular strips. Phys. Rev. B 71(13), Art. no. 134517 (2005)
24. J. Bardeen, L.N. Cooper, J.R. Schrieffer, Theory of superconductivity. Physical Review 108(5), 1175–1204 (1957)

# Chapter 10
# Conclusion and Future Work

## 10.1 Conclusion

This thesis presents a novel study on Second Generation High Temperature Superconductors, which includes their electromagnetic characteristics and AC loss analysis.

Superconducting magnet is one of the most crucial superconducting applications that have been commercially used in medical imaging devices such as MRIs. Lorentz Force Electrical Impedance Tomography (LFEIT) has shown its significant advantages over conventional medical imaging techniques in terms of excellent cancer detection as well as internal haemorrhage, high spatial resolution, portability for emergency diagnostics, and low manufacturing cost. The superconducting magnet is an important component of a LFEIT system, and the combination of a superconducting magnet with a LFEIT system is a sensible approach because superconducting magnets are able to produce magnetic fields with high intensity, which could significantly enhance the SNR of a LFEIT system and enhance the quality of biological imaging. The modelling and simulation of four magnet designs (two based on permanent magnets, and two based on HTS) were carried out using COMSOL Multiphysics. The Superconducting Halbach Array magnet design is able to overcome the disadvantages experienced with other designs. A thin superconducting Halbach Array magnet can achieve the portability (for installation in a general "Type B" ambulance). More importantly, a Superconducting Halbach Array magnet is able to establish a good magnetic field with proper strength and homogeneity for a LFEIT system.

The optimization study of superconducting Halbach Array magnet was carried out on the FEM platform of COMSOL Multiphysics, which consists of 2D models using the *H*-formulation based on the *B*-dependent critical current density and bulk approximation. Optimization of the magnetic homogeneity was executed to increase the number of coils for Halbach arrangement while shrinking each coil's size, but

B. Shen, *Study of Second Generation High Temperature Superconductors:
Electromagnetic Characteristics and AC Loss Analysis*, Springer Theses,
https://doi.org/10.1007/978-3-030-58058-2_10

still maintaining the total amount of superconducting material. The mathematical formulas between the inhomogeneity and increasing numbers of the coils has been derived, which reveals that the magnetic homogeneity can be improved by increasing the number of coils, but the efficiency for homogeneity improvement will decay if the number of coils increases to a high value. Using this optimization method, a Halbach Array configuration based superconducting magnet can potentially generate a uniform magnetic field over 1 T with an inhomogeneity to the ppm level. However, for a real design, the optimization efficiency and fabrication difficulty are required to be taken into account.

The mathematical model for a LFEIT system was built based on the physical principle of the magneto-acousto-electric effect. The magnetic field properties from each magnet design were imported into the LFEIT model which coupled with the ultrasound module from Matlab. The LFEIT model simulated two samples, located in three different magnetic fields with changing magnetic strength and uniformity. According to the simulation result, both improving the uniformity of magnetic field and increasing the magnetic field strength can improve quality of electrical signal imaging of LFEIT. By contrast, increasing the intensity of the magnetic field is more efficient, particularly when the electrical signal generated from the sample is lower than or comparable to the noise level. The simulation results revealed that the quality of signal imaging is still acceptable when the static magnetic field used in the LFEIT system has greater than 10% inhomogeneity, with a magnetic flux density around 1 T. The tolerance of magnetic field inhomogeneity for a LFEIT system is several orders higher than that of MRI, since LFEIT shares the characteristic of ultrasound imaging.

Although there are no actual alternating currents involved in the DC superconducting magnets described above, they experience power dissipation as they are exposed to the varying magnetic field when high current superconducting coils and cables are used in magnet applications, e.g. magnet ramping. This problem generally falls under the classification of "AC loss", which is substantially the same as the problem encountered under AC conditions. Moreover, the DC superconducting magnets for both MRIs and LFEITs also suffer various types of external AC signal disturbances during their operation. Therefore the AC loss characteristics of HTS tapes and coils are still important for HTS magnet designs. This thesis began with the AC loss study of HTS tapes. The investigation and comparison of AC losses on Surround Copper Stabilizer (SCS) Tape and Stabilizer-free (SF) Tape were carried out, which includes the AC loss measurement using the electrical method, as well as the real geometry and multi-layer HTS tape simulations using 2D $H$-formulation by COMSOL Multiphysics. The results demonstrated that hysteresis AC losses in the superconducting layer were frequency independent. The eddy-current AC losses (in Watt) were proportional to the second power of the frequency. The experimental and simulation results revealed that the eddy-current AC losses had almost no effect on the total AC losses for both SCS Tape and SF Tape with transport current frequencies at 10 and 100 Hz, but the eddy-current AC losses started to affect the total losses for the SCS Tape with a transport current frequency at 1000 Hz. Further measurements and simulations proved that the eddy-current AC losses in the copper stabilizer should also be taken into account for high frequency applications over kilo Hz level.

A comprehensive study on AC losses in a circular HTS double pancake coil was carried out using both experiment and simulation. The AC losses from a 2 × 18 circular double pancake coil were measured using the electrical method. In order to improve the consistency with the real circular coil used in the experiment, a 2D axisymmetric *H*-formulation model was established using the FEM package in COMSOL Multiphysics. There are three scenarios which can cause AC losses in an HTS coil: (1) AC transport current & DC magnetic field, (2) DC transport current & AC magnetic field, and (3) AC transport current & AC magnetic field. Moreover, differences in the orientation angle $\theta$ that HTS coil under the magnetic field has been studied for each scenario. For scenario (3), the impact of relative phase difference between the AC current and the AC field on the total AC loss of the coil was analysed. Both the experiment and simulation were carried out for Scenario (1), and good consistency between experimental and simulation results was presented for AC loss magnitude, tendency, and angular dependence. The simulation results for Scenario (2) and Scenario (3) were reasonable and consistent with previous work presented in the literature. Therefore, we believe that the coil model is capable of producing convincing results throughout this multi-scenario study. A powerful coil model has the potential to efficiently compute the AC loss under various complex conditions. In short, a systematic current/field/angle/phase dependent AC loss ($I$, $B$, $\theta$, $\Delta\varphi$) study of circular HTS coil has been completed, and the methods and results of this study will be beneficial for future design and analysis for HTS AC systems.

AC losses of horizontally parallel HTS tapes were investigated by experiment and simulation. Three-parallel-tape configuration was used as an example. The middle tape had less AC losses than an isolated individual tape due to the reduction of the perpendicular field component at its edges, while the end tape had larger AC losses than the isolated tape due to the perpendicular field component at one of its edges. The interactive effect on AC losses became less significant when the gap distance increased, and the loss value started to approach that of the isolated tape. Another idea was sparked that the structure of horizontally parallel HTS tapes can be used for wireless power transmission (WPT). The simulation results showed that a receiving coil with 10 mm diameter cross-section was able to collect a comparable electromagnetic induction (per unit-length) generated by either three parallel tapes or by three turns of HTS coil, with the same AC transport current and the same position of the receiving coil. However, the total AC losses from the three parallel tapes case was only 1/5 of the total AC losses from the three turns coil case. This design has the drawback that the electromagnetic induction decays greatly if the centre of the receiving coil moves far away from the level of the transmitting system, and the conventional vertically parallel coils achieve better wireless power transfer efficiency in a more spacious region. However, the configuration of multiple horizontally parallel tapes in a WTP system has the obvious advantage of lower total AC losses than multiple tapes closely stacked in a conventional coil.

The total average losses per tape of horizontally parallel tapes have been studied. A new parameter $N_s$ has been defined based on the analysis, which is a number of parallel tapes to divide Stage 1 and Stage 2. In Stage 1, $N < N_s$, parallel tapes contributed more interaction to each other, which caused large AC losses in the end tapes that dominate the total losses. Therefore, the total average losses per tape increased with the increasing number of tapes. In Stage 2, $N > N_s$, the far tapes had much less impact on the total losses as the total planar length of the tape system became longer, and the AC losses contributed from the end tapes tended to become relatively negligible. Therefore, the total average losses per tape started to decrease with the increasing number of tapes. The turning point $N_s$, could be a useful factor to determine where the total average losses per tape start to decrease, which will be beneficial for designing an AC system of horizontally parallel tapes. The geometry of horizontally parallel tapes could be potentially applied to superconducting devices like HTS transformers, because this structure is able to maintain and even reduce the total average AC losses per tape with a large number of tapes.

## 10.2   Future Work

Some work can be carried out in future to improve the current study.

For a LFEIT system, experiments can be set up to include: (1) experiment using permanent magnets for a small or middle scale LFEIT system, and superconducting magnets for a large scale LFEIT system; (2) experiment to set up a linear ultrasound transducer fed by a function generator circuit; (3) experiment for a data acquisition module that measures the voltage or currents at the boundary electrodes for the image reconstruction process.

For AC loss analysis, more experiments and simulations can be performed. Various kinds of HTS high current cables can be designed for the use in superconducting magnets, such as the twisted-stacked-tape-cables (TSTC) and conductor-on-round-core (CORC) cables. Detailed AC loss analysis of these high current cables subject to different background fields could be fulfilled. The fully 3D modelling of HTS high current cables could be explored, which would provide a very important and powerful tool for the design and optimization of HTS applications.

Printed in the United States
by Baker & Taylor Publisher Services